PrOTect IT All

A Plain-Spoken Cybersecurity Guide to Protecting Yourself, Your Family, Your Business, and the Systems We All Depend On

Aaron Crow

Host of the PrOTect IT All Podcast

PrOTect IT All

Stories and insights drawn from guests of the PrOTect IT All podcast are retold in the author's own words and paraphrased. No direct quotations are attributed to named individuals.

Published by Aaron Crow

www.protectitallpod.com

Hardcover ISBN: 979-8-9967289-0-9

Paperback ISBN: 979-8-9967289-2-3

Ebook ISBN: 979-8-9967289-1-6

Audiobook ISBN: 979-8-9967289-3-0

First edition, 2026

For my amazing wife, Kary, who hears me on every call and every podcast, knows more technical words than she ever should, and puts up with my crazy schedule, my constant tinkering with nerd gear, and my never-ending hours on the computer.

And for my incredible kids, who wear my swag, think I am a superhero celebrity just because I have a podcast, and somehow still want to hang out with me. Charlie, Sydney, and Tyler, I love you all.

And to the many friends and mentors along the way, I would not be where I am without your guidance.

Contents

Preface

I never set out to write a book. I host a podcast, I do the work, and for years that felt like plenty. But the same thing kept happening. I would finish a talk, or wrap up an episode, or I would just be standing in my own kitchen, and somebody who is not in this world at all, my wife, my in-laws, a buddy from down the street, would ask me a question that stopped me cold with how simple and how important it was. Am I a target? Is my kid safe on that phone? What do I actually do? Those questions are why this book exists.

I did not come to this from behind a desk. The back half of my career has been spent inside critical infrastructure, the power plants and water systems and pipelines that keep the lights on and the water clean, and before that I came up through security and technology the long way. I have been a contractor, an asset owner, a consultant, and a CTO. I have earned hard hats, walked plant floors at shift change, and gotten the two in the morning phone call when something broke. I am telling you that not to brag, but so you know that when I say something works, it is because I have watched it work, and when I say something is a bad idea, it is usually because I have watched that one blow up too.

Here is the thing I have learned. Cybersecurity is not a power plant problem or a Fortune 500 problem. It is a kitchen table problem. The same mistakes that take down a factory are the ones that drain a retiree's savings or pull a teenager into something they cannot get out of. The technology changes, but the people part does not. And the people part is where almost all of this lives.

So who is this book for? If you have a phone and a family, it is for you. If you run security for a big company, there is plenty in here for you too, and I would bet some of these stories sound familiar. I wrote it for the business leader who has to make a call and does not want a lecture. For the IT person who is tired and outnumbered. And for the regular person who just wants to keep their family safe and has no idea where to start. You do not need a technical background. I promise to define things as we go, in plain English, the way I would explain it to a friend over coffee.

This is not a technical manual, and it is not a sales pitch. I am not trying to scare you, and I am not selling you a magic box. I hate fear for the sake of fear, and I hate it when somebody tells you that one product will fix everything. It will not. There is no silver bullet. What there is, is a handful of good habits and a way of thinking that, done over and over, makes you and your business a whole lot harder to hit. People over tools. Action over a perfect plan. Do what you can with what you have.

You are also going to meet a lot of people in these pages. Over more than a hundred episodes of the PrOTect IT All podcast, I have sat down with ethical hackers, CISOs, researchers, plant operators, and government folks who do this for a living. Their stories and their hard-won lessons are woven all through this book, in their own words. I did not learn any of this alone, and you should not have to either.

You do not have to read this front to back. If you want to protect yourself and your family right now, start at the very next chapter and go do the things in it this weekend. If you want to understand how we got here and how attacks really work, that is here too. And if you live in my world of OT and critical infrastructure, the back half of this book was written for you. Read it in whatever order helps you most. Just read it with a pen, because the whole point is to actually do something with it.

One more thing before we start. This whole thing, the podcast, the talks, and now this book, comes from a simple belief. In cybersecurity, everyone has a role to play. You do not have to be the expert. You just have to take the next step. And when you reach the very end of the book, you will find a short invitation from me called Join the Conversation. I mean every word of it, so please do not skip it. So thank you for being here. Let's get to work, and let's protect it all.

About the Author

I am Aaron Crow. I have spent more than thirty years in technology, a big chunk of it in security, and the last half of my career focused on critical infrastructure and operational technology, which is the world of power plants, water utilities, pipelines, and manufacturing.

I came by it honestly. My dad was a controls engineer who built power plants, so I grew up around this stuff before anybody called it OT. I started out in IT, then found my way into the OT world at a Texas utility, where I was responsible for the cybersecurity of more than forty power plants spread across the state, everything from gas and coal to renewable and nuclear. Some of those sites were a six hour drive from the office, and you learn a lot about a problem, and about yourself, on a drive like that.

Since then I have worked across consulting and product companies. I built an OT security practice at a Big Four firm, served as a CTO, and I now advise organizations on protecting the systems the rest of the world only notices when they stop working. Somewhere in there I started the PrOTect IT All podcast, because I kept having conversations that were too good to keep to myself.

Outside of work I am a Texan, a husband to my wife Kary, and a dad to three kids who have heard more about firewalls and network segmentation at the dinner table than any kids probably should. I ruck, I work on cars, I train jiu-jitsu when I can, and I believe, all the way down, that all business is a people business.

If there is one idea I want you to take from me, it is this. You do not have to be the smartest person in the room to make a difference. You just have to care, take action, and bring people along with you. That is the whole game.

Part I

Start Here: Protecting Yourself and Your Family

Chapter One
Cybersecurity Starts at Home

Protecting Your Family, Your Kids, and Yourself

Before we get into threat actors and power plants and all the stuff I geek out about for a living, I want to start with you. Not your company. You. Your phone, your family, and that router sitting in the closet with a password printed on a sticker on the back that you have never once changed.

I have spent my whole career in this stuff, most of it keeping big industrial operations running. But you know what people actually ask me when I am standing at a barbecue with a burger in my hand? Nobody asks me about Stuxnet. They ask, Aaron, am I really a target? What do I actually need to do? And what about my kids?

So here is the honest answer. Yes, you are a target. Not because some hacker picked you out of a lineup. Because nobody has to. Most attacks today are automated. Bots take stolen passwords and try them on millions of accounts while the attacker is asleep. Scam texts go out by the million. The phishing email does not know your name and does not care. You do not get hit because you are special. You get hit because you are reachable, and because something you are doing makes you easy.

And that is actually good news, because the opposite is just as true. Make yourself a little harder to hit and most of that automated junk rolls right past you to the next house with the door unlocked. The vast majority of attacks come down to a person making a simple mistake.

That number sounds depressing until you flip it over. If people are the problem, then people are also the fix. Almost everything in this chapter is just removing the small mistakes the bad guys are counting on. You do not need to be technical. You do not need to spend much. You need about an afternoon and the willingness to actually do it. Because hope is not a strategy.

Start With the Two Keys to Everything: Your Email and Your Phone

If you fix only two things in this whole chapter, fix these. Think about what happens the next time you forget a password. You click forgot password, and where does the reset go? Your email. Or a code gets texted to your phone. That means your email account and your phone number are the master keys to your entire life. Your bank, your retirement account, your kid's school portal, your social media, all of it.

So look at it from the attacker's side. He does not need to break into your bank. That part is hard. He just needs your email. He clicks forgot password on your bank, your bank cheerfully sends the reset link to your inbox, and if he is already sitting in your inbox, he just walks through the front door you left open. Same story with your phone number.

- **Protect your email like it is the front door, because it is.** Give it your longest, most unique password and turn on multifactor authentication, which we will get to in a minute. It is the one account that can reset all the others, so it gets your best lock.
- **Lock down your phone number against SIM swapping.** This is one most people have never heard of. An attacker calls your cell carrier, sweet-talks them into moving your number over to his phone, and then he catches all your text-message codes. Call your carrier and add a port-out PIN or an account passcode. Five minutes on the phone shuts down one of the nastiest attacks on regular people.

Passwords: Stop Reusing, Start Randomizing

If you do nothing else, do this. The most common mistake I see, by a mile, is people using the same password everywhere. And I get it. Who can remember a hundred different passwords? But here is what actually happens when you reuse one.

Some website you signed up for years ago and forgot about gets breached. Your email and your password end up on a list that criminals buy and sell for pennies. And then a bot grabs that email and that password and tries it on your bank, your shopping account, your email, your payment app, and your kid's gaming account. And if you reused that password, boom, they are in. They never hacked a thing. You handed it to them years ago and forgot all about it.

The fix is three words. Long, random, unique. Every account gets its own password, and those passwords should be random junk, not your dog's name with an exclamation point on the end. And I can already hear you. Aaron, I cannot remember a hundred random passwords. Right. That is exactly the point.

- **Use a password manager.** A good one, a reputable and well reviewed password manager, makes up and stores a different random password for every account you have. You remember one strong password. It remembers the other ninety-nine and types them in for you. Honestly, it is easier than what you are doing now, not harder. And I can hear the next question too. What happens if the password manager itself gets breached? Fair question. The good ones are built so that even the company that makes them cannot read your passwords, and your one strong master password plus multifactor keeps the vault sealed. Is that zero risk? No. Nothing is. But reusing one password across fifty sites is a breach already in progress. I will take the vault over that bet every single day.
- **Fix the important ones first.** You do not have to redo everything tonight. Start with your email, your bank, your phone carrier, and the password manager itself. Then chip away at the rest over a few weeks.
- **Assume your stuff is already out there.** Because it probably is. Unique passwords are what keep one old leak from unlocking your whole life.

Turn On MFA Everywhere, and Get Yourself a Hardware Key

Let me define this one in plain English, because the acronym scares people off for no reason. A password is something you know. Multifactor authentication, or MFA, just adds something you have. So even if a criminal steals your password, he is stuck at the door without that second thing. Turn it on everywhere it is offered. Email, bank, social media, shopping, all the things.

- **An app beats a text message.** Codes by text are way better than nothing, but they can get stolen through that SIM swap trick I just mentioned. An authenticator app on your phone is stronger and just as easy.
- **For the accounts that matter, get a hardware key.** A hardware security key is a little device that plugs into your computer or taps against your phone. It is the gold standard because it is basically

impossible to fake. A phishing site can fool you, but it cannot fool the key. Put one on your email, your bank, and your password manager. For about the price of a nice dinner you get the same protection the pros use.

- One more thing while you are in there. When you turn on MFA, most services hand you a set of backup codes, one-time recovery codes for the day you lose your phone. Save them. Print them and put them with your important papers, or store them in your password manager. Locking yourself out of your own account is the most common way people get burned by MFA, and it is completely avoidable.

Lock Down Your Devices

Your phones and laptops and tablets are where your life actually lives now. A handful of settings go a long way.

- **Turn on automatic updates.** Most break-ins use a hole that already had a fix available. The victim just never installed it. Auto-updates close those holes for you while you sleep.
- **Lock the screen and turn on encryption.** A passcode or your face or fingerprint, plus encryption, which is on by default on modern phones and just needs flipping on for laptops. Now a lost or stolen device does not hand over your whole life with it.
- **Control your apps.** Delete the ones you do not use, and check what the rest can get at. If a flashlight app is asking for your location, your microphone, and your contacts, stop and ask yourself why a flashlight needs to know where you are. And only install apps from the official stores.
- **Be careful what you plug in.** Cheap cables and found USB drives can carry malware. Do not plug something you do not trust into a device you do.
- **Back up your stuff.** Photos, documents, tax records. Keep one copy in the cloud and one on a drive you unplug. Think about how people used to keep some photos in a safe in case the house burned down. Same idea. A backup is your undo button for ransomware, for theft, and for the plain bad day when a device just dies.

> *If something is not built with security in mind from the start, you cannot just bolt it on later and expect it to hold.* – A penetration tester and red team leader, on the show

Do Not Trust the Box Your Provider Gave You

That router or gateway your internet company dropped off is the front door to your whole house, and it usually ships wide open. Default admin password, old software, settings tuned for easy setup instead of safety. Treat it like the door it is.

- **Change the admin password on the router.** The default is often printed right there on the sticker, and those defaults are published online for anybody to look up. Change it.
- **Keep it updated, or buy your own.** Update the firmware, and turn off remote administration and any features you never use. A lot of people end up with better security and a lot less junk by buying their own router instead of renting the provider's box.
- **Lock your Wi-Fi.** Use WPA3 or WPA2 and a long passphrase. And do not name your network after your home address or your last name.
- **Put the smart junk on its own network.** Your smart TV, your video doorbell, your thermostat, and your kid's game console are all computers too, and usually the least secure ones in the house. Give them their own guest network. Your smart fridge does not need to sit on the same network as the laptop with your taxes on it. In my world we call that segmentation. In your house it is just a second Wi-Fi network, and most routers set it up in about four taps.

Be Careful What You Hand the AI

AI is in everything now. Your phone, your email, your browser, the little chat box on every website. And it is genuinely useful. I use it every day. But I watch people hand these tools the keys to their entire life without thinking twice, the same way they hand out reused passwords. So let me give you the one rule that keeps you out of trouble. Treat AI like a stranger.

Think about it. Would you walk up to a random person on the street and give them your Social Security number? Your bank login? Your kid's school schedule and your home address? Of course not. So do not paste that stuff into an AI either. Anything you type into one of these tools, you should assume could end up somewhere you do not control. The way I say it on the show, do not put anything into it that you would not be comfortable posting on social media.

There are two ways this burns you at home. The first is obvious. You give it too much, and it leaks or mishandles something it never should have had in the first place. The second one most people never think about. The AI itself can become the attack, tricked by hidden instructions into doing an attacker's bidding. We have already seen real versions of this happen.

So here is how you use AI at home without getting burned.

- **Do not give it the run of the house.** Do not connect it to your whole email account, your entire cloud drive, or anything with money in it. If a tool insists on access to everything, that is a red flag, the same as an app that wants every permission on your phone.
- **Give it a sandbox.** Make one dedicated folder, put only the things you actually want it to see in there, and point the AI at that folder instead of your whole drive. Set up a separate email account for AI sign-ups and experiments, so when something gets messy, and it will, it is not your real inbox.
- **Make it ask permission.** Use the settings that make the tool check with you before it does anything you cannot undo, like sending an email, deleting a file, or making a purchase. Keep a human in the loop for anything that really matters.
- **Never hand it your secrets.** No passwords, no bank details, no Social Security numbers, no private personal information. Not in the chat, not in a file you upload, not anywhere.
- **If you get technical with it, use keys that expire.** If you start wiring AI into your other accounts with what are called API keys, the little passwords that let two programs talk to each other, use ones that rotate and expire, and give the narrowest access that still works. Do not hand it a permanent master key. One of my guests builds guardrails for exactly this for a living, and his whole philosophy is to starve these tools of standing access so that even on their worst day they cannot do much harm.

Protecting Your Kids

This is the hardest part of this chapter to write, and it is the most important. When we hand a kid a phone, we also hand the whole internet a door to that kid. And some of the people on the other side of that door are not selling anything. They are hunting. Predators, sextortion rings, scams aimed right at teenagers. This is real, it is organized, and it is growing. You cannot hand this off to a setting, but the technology does help.

- **Know the devices and the apps.** Use the parental controls built into phones, game consoles, and your router. Screen time, app approval, content filters. Watch for hidden vault apps that disguise themselves as something boring, and for chat apps where a stranger can slide in.
- **Filter at the network and on the device.** Block the known-bad and adult sites for the whole house, which a lot of routers and free family DNS services will do for you, and keep the age settings turned on in the app store and on streaming.
- **Watch in a way that fits their age.** Little kids need close supervision. Teenagers need conversation more than they need surveillance. Be straight with them that you are watching out for them, not spying to bust them.

And here is the control that beats every setting on that list. Keep the conversation open. A predator's best weapon is a kid's silence, and the next chapter covers exactly how to take that weapon away.

And one more thing. Teach your kids the same basics you just learned for yourself. Do not reuse passwords. Do not click links from strangers. And remember that online, anybody can pretend to be anybody.

The Con Is the Real Threat

Most families never get hacked in the movie sense. They get conned. The attacker does not break your encryption. He talks you into handing over the money or the access yourself. The jury duty call from the fake police officer. The tech support pop-up. The romance that suddenly needs a little money wired over. The panicked voice of a grandkid in trouble, which can now be faked with AI for almost nothing. We are going to spend the whole next chapter on this, because it matters that much. For now, just learn the smell of it. Urgency, secrecy, and a strange way to pay. And hold on to the oldest rule in the book. If it seems too good to be true, it probably is, so start asking questions. It gives you every tell and the one rule that beats almost all of them.

The Bottom Line

Almost everything that hits a regular person comes down to a few open doors. A password you reused. An account with no second lock. A device that never updates. A router still using the password off the sticker. A kid with nobody to tell. An AI assistant you handed the keys to your whole life. Close those doors and the automated attacks and the lazy scammers move on to an easier target. You will not be perfect, and you do not need to be. You just need to stop being easy.

Tools That Help

You do not need to spend a fortune, and you do not need any one specific brand. Here are the categories that do the heavy lifting, and what to look for in each.

- **A password manager.** Makes up and stores a different random password for every account. Look for one that is well reviewed, has been around a while, and works on all your devices.
- **Multifactor authentication, an app or a hardware key.** An authenticator app gives you rotating codes. A hardware security key, a little device you plug in or tap, is the strongest option because a fake website cannot fool it. Put the strongest factor on your email, your bank, and your password manager.
- **Automatic updates and built-in device encryption.** Already sitting in your phones and computers. You just have to turn them on.
- **A router you control, with a guest network.** Either your provider's box with the admin password changed and the firmware updated, or one you buy yourself. The feature that matters is the ability to run a separate network for your smart devices.
- **Content and parental filtering.** A family-focused filtering service at the router, plus the parental controls already built into phones and game consoles, to block bad sites and manage what your kids can reach.

- **Backups, one in the cloud and one you can unplug.** Automatic cloud backup for convenience, and an external drive you disconnect so ransomware cannot reach it.
- **AI assistants used with guardrails.** If you use AI at home, lean toward tools and settings that let you control permissions, require your approval before they act, and limit what they can see. Treat broad access as the exception, never the default.

Your Action Steps

You do not have to do all of this today. Knock out these eight this weekend and you have already jumped past most households in the country.

1. **Install a password manager**, and move your email, bank, and phone carrier accounts into it with new random passwords.
2. **Turn on MFA** for your email, bank, and password manager, and order a hardware security key for those three.
3. **Call your cell carrier** and add a port-out PIN to block SIM swapping.
4. **Turn on automatic updates** on every phone, tablet, and computer in the house.
5. **Change your router's admin password**, update it, and set up a separate Wi-Fi network for the smart devices.
6. **Turn on parental controls**, and have the real conversation with your kids.
7. **Put your AI tools on a short leash.** Give them one limited folder, a separate email, and no access to your money or your passwords.
8. **Back up your important stuff**, with one copy you can unplug.

You Will Never Be Perfect, and That Is Fine

Nobody is a hundred percent secure. Anybody who promises you that is selling something, and I am not here to sell you anything. The goal is not perfection. The goal is to be a hard enough target that the automated attacks and the lazy scammers give up and go find somebody easier. Every step in this chapter raises that bar a little. Stack them together and they raise it a lot.

Cybersecurity is not something that only happens at work, to other people, in some industry you have never thought about. It happens at your kitchen table, on your kid's phone, through the box in your closet. The same thing I tell the operators at a power plant is true for you at home. This is not about fear. It is about a few good habits done over and over. So pick one thing off that list and go do it this week. Be the hero of your own story. It starts right here.

Chapter Two

Scams and Social Engineering

How the Cons Really Work

Let me tell you about a phone call. It comes from a local number, and the person on the other end is calm and official. He says he is with the county sheriff's office, gives you a badge number, and tells you that you missed jury duty. There is a warrant out for your arrest. But, he says, there is a way to take care of this today, right now, before a deputy shows up at your door. You just have to pay the fine. And it has to be paid a certain way, and you cannot tell anyone, because this is an active legal matter. Your heart is pounding. You are already reaching for your wallet.

That exact call has hit thousands of people. Smart people. Careful people. Doctors, teachers, engineers, retired cops. And here is what I want you to notice about it. Nobody hacked anything. There was no virus, no breach, no genius typing in a dark room. The attacker did not break into a computer. He broke into you. Your fear, your respect for authority, your need to make a scary thing go away. That is social engineering, and it is how most regular people actually get got.

It Is Not Your Computer They Are After. It Is You.

Social engineering is just a fancy term for conning a person instead of cracking a machine. And attackers love it, because the person is almost always the easiest way in. Why pick a digital lock when you can talk somebody into opening the door for you? They pull on the oldest levers we have. Fear. Urgency. Authority. Greed. Love. You can own every security product on the market and still wire your life savings to a stranger because he sounded official and you were scared.

This is true at home, and it is true at work. The same con that takes a retiree's savings shows up in the office as an email from the CEO to somebody in accounting, asking them to quietly buy a stack of gift cards or push through a wire transfer before the end of the day. Same trick. Bigger check. We even have a fancy name for the corporate version, business email compromise, and it has cost companies billions of dollars. But strip away the suit and it is the jury duty call all over again.

Why Smart People Fall for It

I want to spend a minute on this, because the single most dangerous belief you can hold is that this only happens to people who are not as sharp as you. That belief is exactly what gets sharp people caught. These cons are not aimed at your intelligence. They are aimed at your wiring.

Think about what happens in your body when that sheriff's deputy says there is a warrant out for you. Your heart rate jumps. Your stomach drops. The calm, logical part of your brain steps back, and the old alarm system takes over. That is not a character flaw. That is biology, and the con artist is counting on it. Every good scam is built to trigger that alarm and then keep you too rushed to let the thinking part catch back up.

They use a handful of pressure points, over and over. Authority, so you do not want to push back. Urgency, so you do not have time to think. Scarcity, the deal or the prize that disappears if you wait. Social proof, everybody else already did it. And fear, plain old fear, that your account, your freedom, or your family is at risk. When you feel two or three of those at once, that is not a coincidence. That is the design. And just knowing the design takes a lot of its power away.

AI Just Made the Con a Lot Better

It used to be you could spot a lot of these by the bad grammar and the weird spelling. Those days are over. Today an attacker can use AI to write a flawless message, in any language, polished and personal, and send a million of them before lunch. Worse, the same tools can clone a voice. A few seconds of audio off a video your kid posted is enough to make the phone ring with what sounds exactly like your kid, crying, saying they are in trouble and need money right now. The fake is good. Really good.

Now, I am not telling you this to make you afraid of your own phone. The guests on my show who do this for a living are clear about it. The tools got better, but the playbook did not change. One of them, a former government cyber operator I hosted, put it in a way that has stuck with me ever since.

> *Our job is not to swallow everything whole, but it is not to doubt everything either.* – *A former government cyber operator I hosted*

That is exactly the right posture. Do not panic, and do not dismiss it. The con is more convincing than it used to be, so we just have to be a little more deliberate about how we check.

When the Fake Is Good Enough to Fool the Pros

Here is where it jumps from your inbox to the boardroom. Not long ago, a finance worker in a global company's Hong Kong office got pulled into a video call with his chief financial officer and a few colleagues he recognized. They told him to move money for a confidential deal. He had been suspicious of the email that set it up, but now he was on a live video call looking at their faces and hearing their voices, so he did what they asked. Every person he recognized on that call was a deepfake. The company was out around twenty-five million dollars.

I bring that up not to scare you, but to drive home a point I cannot make often enough. Your eyes and your ears are no longer proof. Seeing is not believing anymore, and neither is hearing. So the check has to come from somewhere the attacker cannot fake. A callback to a number you already trust. A code word. A second channel that you control. The same technology that broke trust cannot be the thing you use to rebuild it.

The Cons, and How to Smell Them

There are a hundred flavors, but they come from the same kitchen. It helps to know that the bad guys reach you through just a few doors. When it comes by email, people call it phishing. When it comes by text message, smishing. When it comes by phone call, vishing. And the newest one, when they hide a bad link inside one of those square QR codes on a flyer or a parking meter, some folks call it quishing. Same con, different doorway. Here are the ones that hit families and businesses the most.

- **Impersonation.** Somebody borrows a uniform you respect. The police, the IRS, Social Security, your bank's fraud department, your boss. The badge is fake, but the fear it triggers is real.
- **The delivery and account-locked text.** Your package is held, click here to reschedule. Your account is locked, log in here to fix it. It is a fake door painted to look exactly like the real one, and it just wants your password.

- **Tech support.** A pop-up screams that your computer is infected and tells you to call a number. The friendly person who answers wants to remote into your machine and, while he is in there, into your bank.
- **Romance, and the slow-burn investment scam.** Weeks or months of attention, and then an emergency that needs a little money, or a once-in-a-lifetime crypto opportunity a new friend wants to share. There is a version of this so patient and so cruel that the criminals themselves named it after fattening up an animal before slaughter. It plays the long game because the long game works.
- **The family emergency.** A grandkid in jail, a child in a car accident, and they need money now, and please do not tell mom. This is the one the voice cloning made dangerous.
- **Fake invoices and vendor changes.** At work, an email that looks like a known supplier saying their bank account changed, please send the next payment here. One quiet digit changed, and the money is gone.
- **Prizes, jobs, and marketplace deals.** A prize you never entered to win. A job that pays great and wants you to deposit a check and send some back. A buyer who overpays and asks for the difference in gift cards.
- The payment-app switcheroo. Somebody overpays you on a payment app and asks you to send back the difference, or a stranger sends money by accident and asks you to kindly return it. The original payment turns out to be stolen or reversed, and the money you sent back was your own. Treat the money apps like cash, because that is how the banks treat them. Once it is sent, it is usually gone.

Every single one of those has the same fingerprints. Once you know the smell, you catch them before they catch you.

- **Urgency.** Act now or else.
- **Secrecy.** Do not tell anyone.
- **Authority.** I am someone you are afraid to say no to.
- **Too good to be true.** Free money, guaranteed returns, a prize out of nowhere.
- **A weird way to pay.** Gift cards, wire transfers, crypto, or a payment kiosk at the grocery store.

Put those together and it is not subtle. If you feel rushed, you are told to keep it a secret, and you are asked to pay in gift cards, you are not in trouble with the law. You are being robbed.

The One Rule That Beats Almost All of It

Here it is. Stop. Slow down. And verify it yourself, on a second channel. Hang up the phone, look up the real number on your own, off the back of your card or the official website, and call them back. Do not call the number the other person gave you, and do not click the link they sent. Go around them. That one habit defeats the vast majority of these cons, because every one of them depends on you acting fast and alone.

No real police officer, no real IRS agent, and no real bank has ever once been mad that you called them back to check. If somebody gets angry or pushy when you say you want to verify, that is not a sign you are being difficult. That is the tell. That is how you know. Slowing down is the whole game. The con runs on speed and pressure and isolation. Take away the speed and the thing falls apart in your hands.

Lock the Doors Behind You

The verify rule stops the con in the moment. These next few moves shrink the damage even when something slips through, and they take an afternoon, not a fortune.

- **Freeze your credit.** This is the big one almost nobody does. A credit freeze is free, you can lift it anytime, and it stops a criminal from opening new accounts in your name even if they have your information. Freeze yours, your spouse's, and your kids', because kids' identities get stolen for years before anybody notices.
- Here is how you actually do it, because it is easier than people think. There are three big credit bureaus, Equifax, Experian, and TransUnion, and you freeze your credit with each one separately, online or over the phone. It is free, by law. Figure ten minutes per bureau. And when you actually need credit yourself, a new card, a car loan, you lift the freeze online in a few minutes and put it right back when you are done. Freezing a kid's credit is a separate process, usually a form and copies of a couple of documents sent by mail, because most kids do not have a

credit file until you create one just to freeze it. A little more hassle, and worth every minute of it.

- **Turn on alerts for every account.** Set your bank and credit cards to text or email you on every transaction, or at least every large one. The faster you see something wrong, the faster you stop it.
- **Use unique passwords and multifactor everywhere,** so a single password phished off one site does not unlock your whole life. We covered this in the last chapter, and it matters just as much here.
- **Keep a separate email for your important accounts.** Use one email address for banking and bills that you never post publicly and never use to sign up for random stuff. It gives attackers a lot less to grab onto.
- **Be a little boring online.** Every detail you post, the pet's name, the high school, the birthday, the vacation, is raw material for the next con. You do not have to disappear. Just stop handing them the script.
- **At work, verify any payment change by phone.** New bank account for a vendor, a wire request from an executive, a change to payroll. Pick up the phone and confirm on a known number before a dollar moves. Make it a policy, not a judgment call.

The First Hour After a Scam

If you realize you just got got, do not spend the first hour beating yourself up. Spend it like this, in this order.

- Call your bank or card company's fraud line first. Wires and payments can sometimes be recalled, but the window is measured in hours, not days. This call outranks everything else.
- Change your passwords second, starting with your email and any account the scammer touched, and turn on multifactor while you are in there.
- Freeze your credit third, at all three bureaus, so whatever they took cannot become new accounts in your name.
- Report it fourth, to the FTC and the FBI's Internet Crime Complaint Center, and tell your family what happened. Silence and shame are what keep these scams in business.

Protect the Two Groups They Hit Hardest

Scammers go after the oldest and the youngest, and for different reasons. Both of them need you in their corner.

Your parents and grandparents. This is not about them being foolish. These cons are good, and they are built specifically to rattle people who grew up trusting a phone call. Have the talk, kindly. Set one simple family rule: any urgent request for money or codes, no matter who it sounds like, you hang up and call me first. And pick a family code word, a goofy word only the family knows, so that when a panicked voice calls asking for help, you can ask for the word. A cloned voice cannot fake what it does not know.

Your kids. Here the danger is different, and heavier. It is not their money the predator is after. It is them. Catfishing, sextortion, a stranger pretending to be a teammate or a cute kid from another school. The technical controls help, but the real protection is a door that is always open. Your kid has to know they can come to you the second something feels wrong, or right after they have already messed up, and that you will help instead of scream. A predator's best weapon is a kid's silence.

A CISO I hosted spends his own time giving these talks to rooms full of parents.

> *If one set of parents hears my talk and it ends up saving a single kid, then the work was worth it.* – *A CISO, on the show*

The Bottom Line

Most people never get hacked in the movie sense. They get conned. The attacker does not beat your technology, he beats your fear and your hurry, and AI has made his fakes good enough to fool your own eyes and ears. You do not stop that with a better gadget. You stop it with one habit, slow down and verify on a second channel, plus a few quiet moves like freezing your credit and turning on alerts that shrink the damage when something slips through. Build those into your family, especially for the kids and the grandparents, and the cons that empty other people's accounts will bounce right off yours.

Tools That Help

You do not need any one specific brand for this. Here are the categories that matter and what each one does for you.

- **Call blocking and spam filtering.** Apps and carrier features that screen out known scam calls and texts before they ever reach you. Your phone and your carrier both offer this. Turn it on.
- **Email spam and phishing filters.** Good email providers catch most of the junk. Keep the filtering on, and report phishing when you see it so the filter learns.
- **A password manager and multifactor authentication.** So a password stolen by a phishing site cannot be reused everywhere, and a stolen password alone is not enough to get in.
- **A credit freeze and identity or credit monitoring.** The freeze blocks new accounts. Monitoring watches your credit and the dark web and tells you when your information shows up where it should not.
- **Transaction alerts from your bank and cards.** Free, built in, and the fastest way to catch fraud while you can still stop it.
- **A family communication plan.** Not software at all. A simple shared rule and a code word for emergencies. It beats every tool on this list for the voice-cloning scam.

Your Action Steps

1. **Adopt the rule out loud as a family:** stop, slow down, and verify on a second channel, every time.
2. **Pick a family code word** for any emergency money request, and tell the kids and the grandparents.
3. **Freeze your credit** and your kids' credit this week. It is free.
4. **Turn on transaction alerts** on every bank account and credit card.
5. **Turn on call and spam filtering** on your phone and through your carrier.
6. **Call your parents** and walk them through the cons in this chapter, gently.

7. **If you ever get hit, report it** to the FTC and the FBI's Internet Crime Complaint Center, and tell people. Silence and shame are what keep these scams in business.

The con artist's whole living depends on catching you alone, scared, and in a hurry. So the fix is almost boring. Slow down, talk to people, and make it normal in your house to check before you act. That is not high tech. It is just human. And it works. Pick one of these and put it into your family this week. Slow down, check first, and you take that whole living away.

Chapter Three

Your Digital Footprint, Identity, and Privacy

What the World Already Knows About You

Give me your full name and the city you live in, and about twenty minutes, and I will hand you back a profile that would make you a little sick. Where you live and what your house looks like from the street. Your spouse and your kids and where they go to school. The cars in your driveway. Your gym, your routine, the coffee shop you tag every single morning. Your mom's maiden name, your first pet, the street you grew up on, which just happen to be the exact answers to your security questions. And I will not break a single law to get any of it. It is all just sitting there, in public, waiting for somebody to come look.

That is not a magic trick. It is called open source intelligence, OSINT for short, and it is the first thing a serious attacker does before they ever come at you. They do not start by attacking. They start by researching. And we have made the research effortless, because we put our whole lives online and forgot that anybody at all can read it.

You Are Already in the Data

Let me get the bad news out of the way first. Your personal information is already out there. Not might be. Is. Over the last decade or so, just about every company you have ever done business with has been breached, and the pieces of you they were holding, your email, your passwords, your account numbers, your Social Security number, got scooped up and sold. There are markets for this stuff, out on the parts of the internet most people never see, where your identity goes for less than a cup of coffee.

And it is not just the big famous breaches. Things leak in dumb, quiet ways all the time. One of my old schools had a backup of its database get loose, a couple of terabytes of it, and the only reason it was a real problem is that the backup was never encrypted, so whatever was in there was simply readable. Nobody ever planned for that backup to become a liability. It just did. Multiply that by every company, school, and app you have ever touched, and you start to get the picture. Every so often, researchers publish another haul of exposed login credentials, and the counts run to the hundreds of millions.

My question for any business owner reading this is simple. Do you even have a way to check whether your people are on a list like that? For most, the honest answer is no. And honestly, that is a fine place to start.

So the goal of this chapter is not to get your data back, because you cannot. That ship has sailed. The goal is to make what is already out there a lot less useful to the people who want to use it against you, and to stop handing them fresh material every single day. You cannot unspill the milk. But you can stop leaving the carton open on the counter.

And if you want one thing to do before you even finish this chapter, go lie on your security questions. The answer to your first pet does not have to be true. It just has to be something only you know. We will get to the rest.

They Research You Before They Hit You

Remember the scams from the last chapter. The reason the good ones feel so personal is that they are. The attacker looked you up first. They know where you bank because you complained about your bank on social media. They know your boss's name because it is right there on a company page. They know you have kids because your feed is full of them. Every one of those details makes the con more believable, and you handed all of it over for free. Treat your public life the way a cop show treats an arrest. Everything you say, or that is simply available about you, can and will be used against you.

Let me show you how powerful public information is, with a story that has nothing to do with crime. I once heard a guy describe how he wanted into a conference that was completely sold out. So he found the one coffee shop right outside the venue, showed up early, and bought every bagel, croissant, and pastry in the place. Bought it all, piled it into a basket. When the attendees came in looking for breakfast and found the shelves empty, there he was, holding a basket. Looking for a bagel? Here you go. Croissant? Right here. And because everybody was wearing their conference badge, he knew exactly who each person was and which ones he actually wanted to meet. He turned a sold-out door into a line of people thanking him, just by reading what was sitting in plain sight. Now picture somebody running that exact same playbook against you, except they are not handing out bagels.

Attackers do this constantly, and they always go for the easy way in, the weakest link, the one nobody is watching. Attackers are like electricity that way. They always find the back door, the easiest path in.

And it is not just people who get exposed this way. It is your stuff too. One of my guests, a researcher, works with a search engine that finds devices sitting wide open on the internet, cameras, control systems, all kinds of things their owners had no idea were even reachable. His whole point is that obscurity is not security. Just because you do not think anybody can see it does not mean they cannot. The same is true of your life. If it is out there, assume somebody can find it.

Your Most Valuable People Are the Target

Here is a question I have actually been asked, sitting in a room full of executives running a practice drill. The person leading it looked around and said, who in here is the one person that, if somebody put a gun to their head, would be the biggest risk to this whole operation? I raised my hand. Because at the time, I knew the passwords to the control systems across all of those power plants. I had accounts on everything. I knew where the firewalls were and how to log into them. If somebody wanted to do real harm, they would not waste time hacking. They would come find me, or somebody like me, and make it personal.

I do not tell you that to be dramatic. I tell you because it reframes how you think about who gets targeted. It is not random. Attackers research, and then they go after the people who hold the keys. In a company, that is your administrators and your executives. At home, that person is you. You hold the keys to your family.

When You Are the Real Target

There is a group of people who get all of this turned up to eleven. Executives, public figures, anybody with money or a name. I have watched companies spend a fortune on physical security for a leader, the guards, the cameras, the cars, and then completely ignore the digital side, where the same person's home address, travel schedule, and family are sitting in plain sight online. That gap is where the real danger lives now. Doxxing, where someone publishes your private details to invite harm. Swatting, where a fake emergency call sends an armed police response to your house. Tracking a target's movements through the data their own devices leak.
The physical world and the digital world stopped being separate a long time ago, and the people protecting the powerful are only now catching up. But here is the thing I want you to take from it. The exact same playbook that protects a CEO scales right down to you. Less exposure. More control. And treating your private information like it matters, because it does.

Your Footprint Is Bigger Than You Think

When most people picture their online presence, they think of their social media accounts. But your real footprint is much wider than that, and a lot of it you never chose to create.

- **What you post.** The vacation photos that quietly announce your house is empty. The new car, the new job, the kid's first day at a school whose name is right there on the sign.
- **The fun little quizzes.** What was your first car, your childhood street, your favorite teacher. Those are not games. Those are your security questions, and you are answering them in public for anyone to collect.
- **The hidden data in your photos.** A lot of pictures carry the exact location they were taken, baked right in. Post enough of them and you have drawn a map of where you are and when.
- **The brokers you never met.** There is a whole industry of companies that quietly buy, bundle, and sell your personal profile, your address history, your relatives, your income range, to anybody willing to pay. You never signed up. You are the product.
- **Your kids' trail.** Every photo, every tag, every account starts a digital record for a person too young to consent to it. Be as careful with their footprint as you are with your own.

And your footprint is not just what you post. It is how you have set up your life. Picture a pretty normal house. The internet provider's box with the default name and the default password, never once changed. The smart speaker, the smart TV, the video doorbell, the smart thermostat, and yes, even the smart fridge and the smart toaster, all sitting on the same network. And on that very same network, the work laptop with sensitive company data on it. That is a lot of doors, and most of them came pre-unlocked. We started fixing that in the first chapter, but I want you to notice here just how much surface you are carrying that you never once thought of as exposure.

Sometimes the exposure is just a choice you did not think all the way through. Somebody has five vacation days, and the last one is a travel day, so they bring the work laptop along to join one meeting from the airport, logging into sensitive systems over open public Wi-Fi because they did not want to burn another day off. I am not saying never do that. I am saying understand what it means when you do. That is the whole game with your footprint. Not living in a bunker, just knowing what you are exposing and deciding on purpose, instead of by accident.

Identity Theft: When Someone Becomes You

Here is where the footprint turns into real pain. When a criminal has enough pieces of you, they do not just log into your accounts. They become you. They open new credit cards in your name. They file a tax return as you and pocket the refund. They get medical care on your insurance, which can quietly poison your own medical records. And the cruelest version, they steal a child's identity and run with it for years, because nobody checks a ten year old's credit until that kid applies for their first car loan and finds out somebody has been busy being them since the third grade.

The nightmare is not the theft itself. It is the cleanup, which can eat months of your life and never quite feel finished. The single strongest move here is the credit freeze from chapter two. If you skipped it then, this chapter is your reason to go back.

Privacy Is a Security Control, Not Just a Preference

Somewhere along the way we started treating privacy like a personality trait. Some people care about it, some do not, no big deal. I want you to think about it differently. Privacy is a security control. Every piece of yourself you keep out of public view is one less tool in an attacker's hands. The smaller your footprint, the smaller your attack surface. That is not paranoia. That is just good defense. Privacy is also a policy fight right now, and I will get into that when we talk about governments and regulation.

The Phone in Your Pocket Is Watching

I cannot end this chapter without talking about the device you are probably holding right now. Your phone is the most effective tracking tool ever built, and you carry it everywhere and pay good money for the privilege. It knows where you sleep, where you work, how long you sit in the school pickup line, and which doctor you visited last Tuesday. And a lot of the apps on it quietly collect that location and sell it, bundled up and passed along to data brokers and advertisers, and from there it can end up just about anywhere.

You do not have to throw the thing in a lake. You just have to take back some of the control you gave away without realizing it.

- **Turn off location for apps that do not need it.** A game does not need to know where you are. Set location to while using the app, or off entirely, for everything that is not a map.
- **Reset and limit your advertising ID.** Your phone hands every app a unique tag that ties all your activity together. Reset it, and turn on the setting that tells apps not to track you across other apps.
- **Trim your app permissions.** Microphone, camera, contacts, location. If an app does not truly need it, do not give it.
- **Be careful with free apps and free Wi-Fi.** If you are not paying for the app, you are often the product. And free public Wi-Fi is a great place for a stranger to watch your traffic, so use a VPN.

The Bottom Line

Your information is already out in the world, and you cannot call it back. But you decide how much you keep adding to the pile, and how useful the existing pile is to the people who want to hurt you. Shrink your footprint. Lock down what new stuff you put out. Freeze your credit so nobody can become you. And treat your privacy as what it actually is, a security control, not a hobby. You will never be invisible, and you do not need to be. You just need to stop being an open book.

Tools That Help

Categories, not brands. Here is what does the work.

- **Credit and identity monitoring. It tells you when your information shows up where it should not, including out on the dark web.**
- **A data broker removal service.** Tools and services that find where your profile is being sold and get it taken down, then keep checking, because it creeps right back.
- **A password manager and multifactor authentication.** So the credentials that leak in a breach cannot be reused to walk straight into your accounts.
- **A privacy-respecting browser and search engine.** Options that collect less about you and block the trackers that follow you around the web.

- **A virtual private network on untrusted networks.** A VPN shields what you are doing on public Wi-Fi at the airport or the coffee shop.
- **Photo and metadata stripping.** Settings and tools that remove the hidden location data from your pictures before you share them.
- **Your social media privacy settings.** The most powerful tool on this list is already built into the accounts you have. Go actually turn the settings up.

Your Action Steps

1. **Look yourself up.** Search your own name, your email, and your phone number, and see what a stranger sees. That is your starting point.
2. Check your email against the known breaches. There are free, reputable breach-notification services that will tell you which breaches your email address has already shown up in, and alert you when it appears in a new one. Two minutes, and it turns I assume I am out there into here is exactly where.
3. **Freeze your credit and your kids' credit, if you have not already done it from chapter two.**
4. **Turn your social media privacy settings up**, and stop doing the quizzes that ask for your security answers.
5. **Get yourself removed from the data brokers**, either by hand or with a removal service.
6. **Strip the location data from your photos** before you post, and share the vacation pictures after you are home, not while you are gone.
7. **Lie on your security questions.** The answer to your first pet does not have to be true. It just has to be something only you know.

None of this is about hiding from the world. It is about deciding, on purpose, how much of yourself you hand out and to whom. Take one of these this week and shrink the target, one choice at a time, and decide for yourself who gets to see you.

Part II

How We Got Here and Who Is Out There

Chapter Four
The Evolution of Cybersecurity

From an Afterthought to Everybody's Problem

There were lumps of coal on my kitchen counter when I was a kid. Actual coal. My sister did a science project built around chunks of it. My dad spent his career as a controls engineer building power plants, the kind of work where you wire a system end to end by hand and label every single connection, and our house was full of the evidence. Programmable controllers on the kitchen table. Talk of units and outages over dinner. I picked this whole world up by osmosis, mostly on fishing trips, sitting quietly while my dad and his coworkers talked shop, long before I had any clue it would become my career, and a long time before anybody thought to call any of it security.

I am telling you that because I have had a front row seat for almost this entire story, and that seat is the reason this chapter matters. The history of cybersecurity is not trivia. Every single problem we are still fighting today traces back to a decision somebody made, or did not make, decades ago. So let me walk you through how we got from a world where nobody locked the doors to a world where a line of code can empty a region's gas stations in a weekend. I lived a good chunk of it, and I will point out where.

When Nobody Locked the Doors

The internet started as a small, trusting place. The early networks that grew into it were built by a handful of researchers who all knew each other, and security simply was not the point. Connectivity was the point. The thinking, to the extent anybody thought about it at all, was that we could always add protection later if it turned out we needed it. That later showed up, and we have been playing catch-up ever since.

Then, in 1988, came the first real wake-up call. A graduate student named Robert Morris wrote a program meant to measure the size of the internet. It had a flaw, and it spread out of control, clogging up something like a tenth of the computers connected at the time. He was not trying to cause harm. But the Morris Worm proved something nobody wanted to admit. The internet was fragile, a single program could spread everywhere in hours, and there was nothing standing in the way. It led to the first coordinated emergency response team, and it should have been the moment we started building security in from the start.

Here is the part that should bother you. That was 1988, and the lesson was sitting right there in the open. Build it wide open, and it will get away from you. We learned it, wrote some papers about it, and then went right on building things wide open for the next thirty-five years.

Crime Discovers the Internet

Through the nineties, computers moved out of the lab and into homes and businesses, and a new kind of trouble showed up with them. The Melissa virus in 1999. The Love Bug in 2000, which hit tens of millions of machines by pretending to be a love letter. The Slammer worm in 2003, which knocked networks offline around the world in a matter of minutes. Annoying, expensive, but mostly chaos for the sake of chaos.

I was getting into the industry right as this shift happened. One of my first real tech jobs was at a computer training company with about ten classrooms, and my job was to wipe and rebuild every single room every weekend so a different week-long course could run clean on Monday morning. One room learning a word processor, the next learning networking software, the next something else. I was learning the plumbing of the entire industry by tearing it down and putting it back together over and over again.

And I watched the threat change underneath me. In the early days it really was mostly mischief, smart people poking at systems for the thrill and the bragging rights. Then somebody figured out you could make real money at it, and that changed everything. Once there was profit in it, the criminals showed up and ran it like a business, because it was one. The pranks turned into an industry, one I will walk you through properly in the next chapter.

When It Got Physical: Stuxnet

For a long time, all of this lived inside computers. Stolen data, crashed servers, money quietly moved around. Bad, but digital. Then, in 2010, the world met Stuxnet, and my corner of this field changed forever. Stuxnet was malware built to do one very specific thing. Reach into an Iranian nuclear facility and physically destroy the centrifuges spinning there, by making them tear themselves apart while the operators' screens calmly reported that everything was normal.

Sit with that for a second, because it is the whole reason my world exists. For the first time, code reached out of the digital realm and broke something physical. An attack made of nothing but software spun a machine until it shattered. The line between the keyboard and the real world, the line everybody assumed was solid, turned out to be more of a suggestion. Just about everything I do for a living traces back to the day that line got crossed.

Meanwhile, On the Plant Floor

Let me back up and tell you where I was while all of this was happening, because the view from the plant floor explains a lot about why we are where we are. Early on, one of my jobs was rolling out servers to power plants so the old green-screen terminals could talk back to the mainframe. I built racks of them, loaded them into the back of a U-Haul, and another guy and I drove them across the country, unloading them by hand and rolling them into what they generously called a server room, which was usually a closet that happened to have power. That was the level of thinking back then. Get it working. Security was not on the list.

Years later I went back to that same world as what we call an asset owner, eventually responsible for more than forty power plants across Texas. And I will never forget a control system vendor pitching us an upgrade. They laid out the options and asked, in these exact words, do you want the secure version or do you want the insecure version? The secure one cost more. We bought it, and then I told them they needed to stop calling it that, and walked them through every reason why. I do not think they call it that anymore. But sit with the fact that, not that long ago, secure was the upcharge.

When I started pushing cybersecurity into those plants, I had almost no budget and a whole lot of resistance. The work was driven by regulation, and I would walk into a plant doing a control upgrade and tell them they had to add a few hundred thousand dollars to do it the secure way, money that came straight out of their maintenance budget. That is not a fun conversation. So I stopped selling cyber and started selling value. Because the same monitoring tools that watch for an attacker also catch the boring operational problems nobody else can see.

Here is what I mean. One of the first plants where we turned on full monitoring had a beautiful, fully redundant design. Redundant switches, two network cards on every device, redundant power, the whole nine yards. The second we plugged our system in, it lit up with an alert. One of the switches, tucked in a cabinet in a back room, was screaming for help, and not one person knew, because nobody had opened that cabinet to see the blinking light. The switch had failed, but the redundancy held, so the plant kept right on running while quietly operating with no backup at all. We walked over, opened the cabinet, and found a power supply cooking itself because a zip tie had fallen into the fan and stopped it from spinning. We pulled the zip tie out, the fan spun up, the switch rebooted, and the redundancy was back. It took us fifteen minutes to find. It had probably been like that for who knows how long, because nobody knew to look.

Another plant had a machine that kept rebooting itself around two in the morning on Saturdays, with nothing in the logs to explain it. So we put a person in a chair right next to the computer to watch. Around two a.m., an operator walked in, said a few words I will not print here, and rebooted it. My guy asked him why. The operator shrugged and said it freaking happens every time during a certain process, the screen freezes, you reboot it, ten minutes, done. He never logged it because to him it was not a problem. It was just part of the job. Everybody who trained him did the exact same thing. It was a memory leak that had been there for years, completely invisible to anyone trying to help from the outside, until we put a human in that chair. I tell you those stories because that is the real history of how defense grew up. Not with some magic product. With people finally turning on the lights and looking.

When It Got Personal: Breaches and Ransomware

While the nation-state stuff grabbed the headlines, a quieter disaster was building underneath. Company after company got breached, and the personal data of ordinary people poured out. The Sony hack in 2014 spilled a studio's emails and secrets for the world to read. The Equifax breach in 2017 exposed the personal information of around one hundred forty-seven million people, one of the most sensitive piles of data in the country, held by a company most of those people never chose to do business with in the first place.

And the breaches taught us something important about how attackers actually operate. Take the big retail breach from about a decade back. The attackers did not smash through the front door, which was reasonably well guarded. They got in through a heating and cooling contractor, a third party with a remote connection into the network. They went after the supplier because the supplier was the soft spot. Attackers do not attack your strength. They find the back door, the path of least resistance, the same way water finds the low ground.

And then ransomware grew up. Instead of just stealing your data, the criminals started locking it up and demanding payment to give it back. Cybercrime stopped being something that happened to faceless corporations. It started hitting the places we all depend on, the ones that cannot afford to be down.

The year the whole world felt it was 2017. Two worms went around the planet within months of each other. The first, the one the world came to know as WannaCry, locked up hundreds of thousands of computers across something like a hundred and fifty countries and hit hospitals in the UK so hard that surgeries got postponed and ambulances got diverted. The second started as a poisoned software update in one country and spilled out to the whole planet, shutting down one of the biggest shipping companies on earth so completely that they were tracking containers with paper and whiteboards. Billions of dollars in damage, and neither worm cared who you were. That was the moment ransomware stopped being a line item on somebody's security budget and became a thing that could stop a surgery and strand a ship. Hold onto that, because it set the stage for what came next.

When IT Took Down a Power Plant

I want to tell you one more story from my own career, because it captures everything this chapter is about. At a multi-unit power plant, a control system vendor was on site during a planned outage, doing an upgrade. One unit was down for the work. The other was up and running and making money. The whole place ran on a flat network, no real walls between the units, and a single shared system for handling logins across all of it.

Part of the job was rebuilding one server. The technician doing it did not have deep experience with that kind of system, and instead of carefully rebuilding the one piece, he rebuilt the whole thing from scratch, giving it the same name and the same address as the old one. To the network, a brand new authority had just shown up claiming to be in charge, and this new one did not recognize any of the existing equipment. The running unit started asking it for permission to do normal things, and it said no to every single request. Hundreds of thousands of denials an hour.

In the control room, the numbers on the operators' screens started to scramble, and then they froze. Big goose eggs, nothing, where the live readings used to be. The unit itself was running perfectly. It was doing exactly what it was supposed to do. But the operators could no longer see it, and an operator who cannot see a unit cannot trust a unit. So they did exactly what they were trained to do. They tripped it offline, on purpose, and refused to bring it back until somebody could explain what had happened. The vendor swore up and down they had not caused it. We came in, pulled the logs, and showed them, line by line, exactly what had happened.

The plant manager did not care one bit that it was an accident instead of a hacker. A unit had come offline, during an outage, and the place was losing money. And here is the part I really want you to hear. That plant probably would have told you, with a straight face, that they had run for forty years without a single cyber incident. They were wrong. They had one. It just did not look like the movies, so nobody recognized it for what it was. How many of those quiet incidents do you think are sitting in plants and offices and homes right now, written off as a fluke because nobody pulled the logs?

When It Shut Down the Country: Colonial Pipeline

In May of 2021, a ransomware attack hit Colonial Pipeline, the company that moves a huge share of the fuel up the East Coast of the United States. Here is the part people forget. The attack never even touched the part of the company that physically moves the fuel. It hit the business side. But the company shut the pipeline down anyway, out of caution, and within days there were gas lines, panic buying, and stations running dry across a big chunk of the country. They ended up paying the attackers millions to get moving again.

People love to correct me on this one. That was not an OT attack, they say, the ransomware hit the business network, not the pipeline controls. And technically, they are right. But they are missing the entire point. The company shut the pipeline down anyway, because they did not have a clear enough understanding of their own risk to know whether they could keep it running safely. The label did not matter. The gas lines were real. Your CEO does not care whether it was IT or OT either. They care about two things, what happened and how do I get it back.

I will give you an even starker example. Picture someone attacking the physical equipment directly, with no keyboard involved at all, and knocking out power to a lot of people anyway. That is not a cyberattack, but it takes down the same systems a hacker would have gone after. An attacker does not care whether the thing they break is labeled IT, or OT, or physical. They care that it breaks. It is all just business risk, and arguing about the label while the lights are off is a waste of breath.

Where We Are Now

Which brings us to today, and a flat-out arms race. Attackers use AI to move faster than ever. Nation-states quietly pre-position inside critical infrastructure and wait. Supply chain attacks poison the software that thousands of companies rely on, so a single break-in turns into ten thousand. We even got a reminder that you do not always need an attacker at all. In 2024, one faulty software update crashed millions of computers around the world in a matter of hours, grounding flights and freezing hospitals, and nobody attacked anything. It was just fragility, on full display for everyone to see. And the real attacks keep coming, right up to recent strikes on power grids that did actual, physical damage to the systems that keep the lights on.

Meanwhile, a lot of what we are defending is old, and getting older. One of my guests, a veteran security practitioner who has spent his career staring at the aging infrastructure underneath all of this, put his finger on exactly why it never seems to get fixed.

> *For a lot of organizations it is just cheaper to pay the parking ticket than to actually fix the problem. That is where we keep missing the point.* *– A veteran security practitioner, on the show*

What the History Actually Teaches Us

So why drag you through forty years of this? Because the same handful of lessons keep showing up, and we keep refusing to learn them.

- **We build first and bolt security on later.** From the early internet to the smart device in your kitchen, we connect it before we secure it, and we pay for that choice down the road. One of my guests, a researcher, spends his days finding gear that got put on the internet years ago with no thought to security at all, still sitting there wide open.
- **The fundamentals never go out of style.** Every era's big disaster traces back to a basic that got skipped. A patch not applied, a network not segmented, a password reused, a backup never tested. The tools change. The fundamentals do not.
- **Underneath the technology, it is always a people and a process story.** Somebody made a decision, or failed to. Somebody got rushed,

or trusted the wrong message on a bad day, or rebuilt the whole server when they meant to rebuild one piece.

- **And it is everybody's problem now.** This started as a niche concern for a few researchers. Today it is your gas, your water, your hospital, your bank, and your family.

I actually find that history hopeful, in a strange way. Because if the problems keep turning out to be the same ones, then so do the answers. We are not waiting on some magic invention to save us. We already know what works. We just have to go do it.

The Bottom Line

Cybersecurity did not arrive overnight. It grew up right alongside the internet, one painful lesson at a time. From a worm nobody meant to release, to a weapon that shattered centrifuges, to a domain controller that left a control room staring at goose eggs, to a ransomware note that emptied gas stations across half a country. Through all of it the pattern holds. We connect things before we secure them, we skip the boring fundamentals, and we treat security as somebody else's job right up until the day it lands on our own doorstep. The story is still being written, and you are part of it now. The good news is that the lessons are clear, and they are the same ones the rest of this book is built on.

Tools That Help

History keeps pointing at the same short list. These are the categories that would have prevented or softened nearly every disaster in this chapter.

- **Patch and update discipline.** Keeping systems current closes the holes that old attacks keep riding in on.
- **Asset inventory.** You cannot protect what you do not know you have, and most victims had no real idea what they were running.
- **Network segmentation.** Limiting how far an intruder, or a mistake, can spread turns a catastrophe into a contained incident. A flat network is how one rebuilt server takes down a whole plant.
- **Monitoring and logging.** You cannot respond to what you cannot see. Half the stories in this chapter were invisible until somebody turned on the lights.

- **Backups you have actually tested.** The reliable answer to ransomware for forty years running, as long as you proved they work before you needed them.
- **Incident response planning.** Because the question was never if, it was when, and the prepared recover in hours while the unprepared recover in weeks.

Your Action Steps

1. **Read the next big breach with new eyes.** When it hits the headlines, ask what fundamental got skipped. It is almost always somewhere on the list above.
2. **Pick one fundamental and check whether you actually do it.** Updates, backups, monitoring, or just knowing what you own. Be honest with yourself.
3. **Test a backup this month.** Restore something and prove it works before the day you are desperate for it.
4. **Assume you have already had a quiet incident.** Go look at the thing everybody writes off as a fluke. It might be your goose eggs.

We keep relearning the same lessons the hard way. You do not have to. That is the whole point of knowing the history. Pick one and put it to work this week.

Chapter Five

The Threat Landscape: Who Are the Attackers?

Forget the Hoodie in the Basement

Picture the hacker in your head. I will bet money it is a young guy in a hoodie, alone in a dark room, lit up by three monitors, fingers flying, breaking into something important for the thrill of it. Hollywood gave you that image, and it is almost entirely wrong. The people coming after you and your business are mostly not lone geniuses chasing a rush. Most of them are running a business, and a wildly profitable one. If you want to defend yourself, you have to actually know who is on the other side of this.

So let me introduce you to the real cast, the way the people who chase these folks for a living describe them. One of my guests, an industrial incident responder, deals with these attacks as a full time job, and the first thing she will tell you is that this is neither rare nor theoretical.

> *Industrial systems are under attack constantly. Responding to it is my full-time job, and my team's too.* – An industrial incident responder, on the show

Cybercrime Is an Industry Now

Here is the single most important thing to understand about the threat landscape. Cybercrime grew up. It is not a hobby anymore. It is an industry, one that costs the world trillions of dollars a year by some counts, with all the trimmings of a real business. There are marketplaces where you buy stolen data and rent attack tools. There are specialists who only do one part of the job and hand it off to the next crew. There is, and I am not exaggerating, customer service. Ransomware became a service you can subscribe to, so somebody with almost no skill can point professional-grade tools at a hospital or a small town and collect.

And the business model keeps evolving. The newer crews do not just lock your data. They steal a copy first and threaten to publish it, which means even a company with perfect backups still has a very bad day coming. We will dig into that whole nightmare, and how you actually beat it, in the ransomware chapter.

That shift is why this touches everyone now. When it was a few people doing it for kicks, you had to be unlucky or important to get hit. When it is an industry optimized for profit, getting hit is just a function of being reachable and being easy. A red team founder who came on the show points out the strange silver lining in all of this. After enough of these attacks made the news, the whole world finally learned the vocabulary.

> *There is one silver lining to Colonial Pipeline. Ransomware became a kitchen-table word. Now even your grandmother knows what it is.* – *A red team founder, on the show*

And do not picture a tidy criminal hierarchy with a mastermind at the top. One of my guests, who has sat across the table from these groups in ransomware negotiations, describes today's underground as pure chaos. The old gangs that had rules, the ones that used to say we do not hit hospitals, are mostly gone. Crews form, break up, and reform constantly. They poach each other's work. There was even a stretch of a couple of years where a lot of the big players went quiet, because the people behind them did not know if they were about to get pulled into a war. The point is, there is no honor among these thieves, and there is no front office to reason with. It is a swarm, not an army.

And the swarm got faster. He told me that when he started in this field, a serious attack meant an intruder who spent a hundred and forty days carefully, quietly mapping a victim's network before making a single move. That patient craftsman is mostly gone. Today the typical attack is smash and grab. Get a handful of people to click, figure out which of them has access to something valuable, and fire off ransomware at whatever you can reach, all in a day or two. The same low-skill crews that used to hit maybe five companies a month now hit a hundred or five hundred, because the tools do the heavy lifting and nobody is being careful. Noisier, clumsier, and a whole lot more of it.

And if you are wondering why somebody does not just arrest them all, here is the ugly truth. Most of these crews operate from countries that will never hand them over, places where the local government looks the other way as long as the attacks point outward, so our investigators can name them and indict them and the crews just keep right on working. The internet does not have borders. Law enforcement does.

The Cast of Characters

Attackers are not all the same, and lumping them together is how you build the wrong defenses. Here is the real lineup.

- **Commodity criminals.** The bulk of what hits a normal person or business. Automated, financially motivated, running ransomware and stealing whatever they can sell. They do not know your name and do not care. They are casting a giant net and pulling in whoever is easy.
- **Script kiddies.** The term sounds harmless, and it used to be. These are people who cannot build anything themselves. All they can do is run a tool somebody smarter made. The catch is that the tools, and now AI, do the hard part for them, so a person who understands almost nothing can still do real damage at scale.
- **Hacktivists.** Not in it for money. They want to make a statement, embarrass a target, or protest something. The damage is real even when the motive is a cause.
- **Insiders.** Your own people. Usually by accident, occasionally on purpose. More on them in a second, because this one surprises people.
- **Nation-states.** The patient, funded, professional end of the spectrum. Spies and saboteurs working a long game. More on them too.

Now here is the part that ties it together, and after thirty years of watching this space I believe it in my bones. These categories are not new. We have always had criminals, insiders, activists, and nation-states. What changed is not the cast. It is the tooling, and how cheap and fast it got. The same actors, with vastly better weapons. The threat intelligence folks I host see it the same way.

A quick word on those script kiddies, because there is a story that circulates in the community, and I have never been able to verify it, but it captures exactly why running tools you do not understand is dangerous to everyone, including the person running them. The story goes that years ago, somebody took a popular attack tool, slipped a hidden change into it, and put the booby-trapped version out where the script kiddies would grab it. Everyone who downloaded it and fired it off quietly reported their own identity to law enforcement, and a wave of arrests followed. The people using it had no clue what the tool was really doing, because they never do. That is the whole definition of a script kiddie. They are holding a loaded weapon and they cannot see the trigger.

The Insider You Did Not Plan For

When people hear insider threat, they picture a spy or a disgruntled employee stealing files on their way out the door. That happens. But the vast majority of insider incidents are just regular people making honest mistakes. Somebody clicks the link. Somebody plugs in the cable. Somebody spins up a workaround because the official way was too slow. And here is a truth that incident responder drove home that most companies do not want to hear. A lot of that risky workaround behavior, what we call shadow IT, is not the employee's fault. It is the security team's fault.

Her point is that people go around the rules when the rules make their job impossible. If security says you cannot do the thing you need to do to keep the plant running, the operator is going to find a way to do it anyway, off the books, out of sight. You did not stop the risky behavior. You just pushed it into the shadows where you cannot see it. The fix for that kind of insider is not more punishment. It is building security that works with people instead of against them.

The intentional insider is rarer, but it is real. The same incident responder has investigated cases where somebody deliberately caused harm, to a system or even to a person, and her team had to piece together who did it and how. And remember the reframe from back in chapter three, the one about who attackers really go after. It is the person who holds the keys. At work, that is your administrators and executives. At home, that is you, and the people you have handed a device to without a conversation.

The Patient Ones: Nation-States

At the far end of the spectrum are the nation-states, and they play a completely different game. They are not smashing and grabbing. They are patient, funded, and professional. In some cases they are after secrets, stealing intellectual property and manufacturing designs. In others, and this is the one that keeps people like me up at night, they are pre-positioning. Quietly working their way into critical infrastructure and just sitting there, mapping it, building access, waiting. So that if a real conflict ever comes, they already have a hand on the switch.

The same incident responder made a point that should reset how you picture these people. The serious adversaries going after industrial systems are not just hackers. They are engineers. They understand the process, the safety controls, and the physics, often as well as the people who run the plant. This is not somebody guessing. This is a professional who did their homework for years. We are still, right now, untangling foothold after foothold that sophisticated adversaries quietly established in systems we all depend on.

I will give you the strangest version of this I have personally run into. You met Stuxnet last chapter, the nation-state weapon built to destroy centrifuges. Well, I have actually seen a copy of it running loose in a power plant. It had spread there somehow, but the specific equipment it was built to attack was not present, so it just sat there, doing nothing, harmless. We cleaned it up and moved on. But sit with that for a second. A military-grade cyber weapon, escaped into the wild, idling quietly inside a power plant, because that is just where the wind blew it. That is the world we actually live in.

Here is the part that makes the nation-state game even harder to defend. These adversaries do their homework in a way you cannot easily match. Because the big vendors standardized their gear, the same models, the same default setups, and the same network layouts show up in plant after plant, so the old comfort of just being obscure and one-of-a-kind is gone. A serious adversary can buy or build a copy of your exact equipment, set it up in a lab, and practice the attack until it works perfectly, long before they ever touch you. They are not improvising on the day. They are rehearsing for it.

The Barrier Just Collapsed

If you take one thing from this chapter, take this. The wall that used to keep most people from being able to attack anything just fell down. It used to take real expertise to understand a complex system well enough to break it. That expertise was the speed limit on how many capable attackers existed. AI just removed the speed limit.

One of my guests, an OT penetration tester I hosted, ran an experiment that I cannot stop thinking about. He took a screenshot of the control logic running an industrial process. He told an AI almost nothing, just what kind of system it was and that it controlled things like temperature. The AI handed back a full breakdown of how the process worked, exactly which alarm values to change, what would happen to the process if you changed them, and then, unprompted, a ready-to-run script to do the whole thing. Then he made the point that should land hard.

> *Once they can get this information, they do not need to be an expert. They could be a high school student.* – An OT penetration tester, on the show

One of those threat intelligence guests showed me the everyday version of the same thing. The old scam email full of bad grammar that you could spot a mile away. Now the scammer pastes it into an AI and says, write this in the voice of a sixty year old from Houston, and out comes something clean and convincing, tested across thousands of targets until the click rate is thousands of times higher than anything we used to see. The script kiddie who only knows the words this is a login page can now ask an AI to write the actual attack. The pool of people who can hurt you did not grow a little. It exploded.

The Movie Hacker Versus the Real One

I told you at the start of this chapter that the hoodie is a myth. Here is the version that kills it all the way dead, because believing it makes you defend the wrong things. In the movies, the climax is a hacker furiously typing until a screen flashes access granted, and then the dam bursts. The people who actually respond to these attacks for a living laugh at that. In the industrial world, taking over one of those little control devices is the easy part. They have almost no security built in. You send one a stop command and it stops, no questions asked. The hard part, the part that takes a real expert, is understanding the process well enough to make something genuinely bad happen without the safety systems catching it first. The dramatic break-in is trivial. The knowledge of what to do next is the actual weapon.

How they actually get from that first click to the machine that matters is the whole next chapter.

Why They Win, and Why You Can Still Beat Them

You will hear this line a lot in security. The defender has to be right every single time, and the attacker only has to be right once. It is true, and it sounds hopeless. But the incident responder I introduced at the top of this chapter has no patience for doom and gloom, and she pushed back on it hard.

> *A lot of these criminal actors will move on. If you are too hard a target and you are not worth the effort, they go find an easier one.* – An industrial incident responder, on the show

That is the whole strategy of this book in one sentence. Remember, most attacks are not personal. They are a giant automated net looking for easy. The thing that gets you caught in it is simply being below a certain bar of difficulty. So you do not have to be an impenetrable fortress. You do not have to beat a nation-state. You just have to be harder than the next target, so the automated tools and the lazy criminals give up and go find somebody easier. Deterrence is real. You can earn it with the fundamentals, the same ones we keep coming back to.

There is one more shift worth knowing, because it changes the stakes on the defending side too. For a long time, security was something an organization could quietly underfund while nobody paid a personal price. That is ending. We are now seeing security leaders held personally, even legally, accountable when a breach is hidden or mishandled. The same threat intelligence guest, who watches this closely, is careful not to judge any one case, but the direction is clear. The era of treating security as somebody else's problem, or as a box you can quietly ignore, is over. The threat got more serious, and so did the responsibility to take it seriously.

The Bottom Line

The attackers are not who the movies told you. They are an industry, organized and profit-driven, plus a long tail of opportunists and a small number of patient professionals working for governments. The cast has not changed much in thirty years, but their tools just got a thousand times better and a thousand times cheaper, so the number of people who can come after you went through the roof. The good news is that most of them are not after you specifically. They are after easy. Stop being easy, and the vast majority of this threat landscape simply rolls past you to somebody who did not read this book.

Tools That Help

Knowing the enemy points you at the right categories of defense.

- **Threat intelligence.** Knowing what is being used and who is being hit, so you are defending against what is actually happening, not what happened ten years ago.

- **Monitoring and detection.** Since you cannot keep everyone out, you need to see them when they get in. This is what turns a quiet, year-long breach into a quick incident.
- **Multifactor authentication.** The single best way to make a stolen password worthless, which knocks out a huge slice of commodity attacks.
- **Network segmentation.** So one foothold cannot become the whole company. It limits what any attacker can reach.
- **Asset inventory.** You cannot defend what you do not know you have, and most victims did not know what they were running.
- **Security awareness training.** Your people are the most targeted layer, so they have to be a trained layer, not a blind spot.
- **Dark-web and credential exposure monitoring.** A way to know when your information, or your employees', shows up for sale, so you can react before someone uses it.

Your Action Steps

1. **Figure out who is most likely to hit you.** For almost everyone reading this, it is automated, opportunistic crime, not a nation-state. Defend for that first.
2. **Decide to be the hard target.** Do the fundamentals so the automated nets pass you by. That is the whole move.
3. **Treat your people as a layer, not a liability.** Train them, and build security that does not force them into the shadows.
4. **Check whether your credentials are already for sale.** If you run a business, get a way to see if your team shows up in the breach data.

You do not have to outrun the bear. You have to be a little faster than the other folks the bear is chasing. Pick one fundamental and go tighten it this week. Keep paying attention, and the odds tip your way.

Chapter Six

How Attacks Actually Work

From the First Click to the Ransom Note

At a conference in Vegas, one of my guests, a penetration tester and red team leader, was holding a phone charging cable. It looked completely normal. It was not. Inside that cable was a tiny computer, a little wireless access point, and even GPS, so it could behave like an ordinary cable at your house and then come alive the moment you plugged it in at the office. A guy he was talking to swore up and down he could spot a fake. So he tossed the cable over his shoulder into the crowd. Within two seconds, a total stranger picked it up, said hey buddy, you dropped your charging cable, and handed it right back. Nobody thinks twice about a cable. Drop one in a parking lot, and somebody assumes their kid left it in the car and plugs it straight in.

I open with that because it tells you almost everything about how real attacks work. You already know from last chapter that the hoodie in the basement is a myth. Here is what replaces it: a cheap cable, a tired employee, a forgotten remote connection, a password somebody reused. In this chapter I am going to walk you through how an attack actually unfolds, start to finish, so that the next time you read a scary headline, you understand what really happened, and just as important, what would have stopped it.

An Attack Is a Process, Not a Moment

The most useful thing you can understand is that an attack is not a single event. It is a process, with stages, and the bad guys move through them more or less in order. First they research you, the open-source intelligence we covered earlier, learning your people, your vendors, your weak spots. Then they get in, which is the foothold, the first compromised account or machine. Then they move, spreading sideways from that first foothold toward the things that are actually worth money. And finally they act, stealing the data, locking it up for ransom, or breaking something. Get in, move, act. Almost every breach you have ever heard about followed that path. And the good news hiding in there is that you get a chance to catch them at every single stage.

The Front Door Is Usually a Person

Forget the firewalls for a minute, because that is not where most attacks begin. They begin with a person. One penetration tester, who breaks into companies for a living, puts a number on it.

> *Eighty to ninety percent of most attacks come down to social engineering.* – *A penetration tester and red team leader, on the show*

When a headline says the hackers broke through the firewalls, the truth underneath is almost always that somebody clicked. Here is a real example you have probably heard about without knowing the details. At one big company everybody uses, attackers already had an employee's password. The only thing standing in their way was that second login approval that pops up on your phone. So they just kept sending it. Over and over, late at night, ping after ping, until the exhausted employee, half asleep and sick of his phone buzzing, finally tapped approve just to make it stop. That one tap was the entire breach. They did not beat the technology. They wore down a tired human.

And the lures keep getting better. The same penetration tester was speaking to a room of students about to graduate, and he showed them a classic clumsy phishing email, the kind with the generic greeting and the slightly-off logo. They all caught it, proud of themselves. Then he showed them a second email, this one about the very conference they were sitting in, with the right dates, the right names, the right details, telling them their registration needed fixing in the next day or two. The room gasped and swore it was real. An AI had written it in seconds, and it even explained why it would work.

The Easiest Way In: Remote Access

When the attacker is not conning a person, the next most common way in is a connection you forgot you had open. One of my guests, an OT threat researcher, leads a research team, and they pulled real attack data from across thousands of industrial networks to find out what attackers actually use. The number one technique was not some exotic zero-day. It was remote access.

> *Remote access, old CVEs, lateral movement. Start from there. That is your starting point.* – *An OT threat researcher, on the show*

Think about how many doors got cut into networks overnight when the world went remote. And think about all the vendors who have a standing connection into your systems right now, because the contract requires it, that you have never once audited. One story from the show captures the vendor version of this. A hospital had a third party's line running straight into its network, with the ability to do basically anything, and zero oversight. When the tester asked if that worried anyone, the answer was, well, nothing has happened yet. That is the same back door behind the big retail breach from chapter four.

I lived the painful version of doing this right. Before there were good tools for it, the only safe way I could give a vendor remote access to a power plant was to chain them through hop after hop. Log into one jump box, then into another, then another, with checks at every step, like the movie Inception where you are a dream inside a dream and you cannot remember which level you are on. The vendors hated it. The operators hated it. But it kept the control network off the open internet, and that was the entire point.

They Use the Holes You Left Open

Here is a finding from that OT threat researcher's data that I want you to really sit with. When his team looked at the vulnerabilities attackers are actually exploiting in the real world, not the theoretical ones, not the scary headline ones, the ones genuinely being used, more than sixty percent of them were five to ten years old. Not zero-days. Not cutting-edge anything. Known holes, with known fixes available, that nobody ever got around to patching.

That is the real story of how attackers get in. Not genius. Persistence against neglect. The default password nobody changed. The old machine that cannot be patched and got left on the network anyway. The system somebody stood up for a project five years ago and forgot. The back door is usually not picked. It is just standing open, the way it has been for years, waiting.

Sometimes the open door is one your own people propped open without thinking. On one assessment I heard about on the show, the testers found a single computer wired to both the open internet and the internal network at the same time, a perfect bridge for an attacker to walk straight across. The kicker is that the person who finally admitted it was his own machine was the same guy who had invited the testers in and sat through their whole presentation. He said, a little sheepishly, I was just testing something. That is how most of these holes get made. Not by villains. By busy people taking a shortcut they fully intended to undo later.

Getting In Is Not Getting Caught: Moving Sideways

Landing on one machine is not the goal. The first thing they compromise is rarely the thing they want. So they move sideways, hunting for the valuable stuff, and a flat network is a wide-open field for that. A red team founder who came on the show tells a story about a client who felt safe because they had firewalls between their office systems and their critical systems. Then he pointed out they shared a single login system across both. His reaction was blunt.

> *With that shared directory, you just made it one level. You flattened your whole network. It is built to go right through those firewalls. That is the point.* – *A red team founder, on the show*

Once they are moving, the smart ones do not bring noisy hacker tools. They live off the land, using the normal built-in tools that are already there, so they blend right in with legitimate activity. One red team leader on the show showed me the one that should make every business owner check their settings. On nearly every test he runs, his team simply maps a network drive straight out through the company firewall to a server they control, and drags the stolen data out the door like they are copying files to a thumb drive. Almost no firewall blocks it, because everybody watches traffic coming in and nobody watches the normal-looking traffic going out. As he put it, everybody opens a map drive through the firewall. An OT penetration tester I hosted adds the scariest flavor of this. The attacker using real, stolen credentials, who looks exactly like a normal employee, is the hardest of all to catch, because nothing they do looks wrong.

To see how reachable the important stuff really is, consider what one ICS security instructor's team does for a living. A couple of months before I talked with him, they tested one of the larger airports in the country, live, while it ran around the clock. They went after three of its most sensitive zones, the baggage handling, the security cameras, and the door access controls, with the lightest possible touch so they would not actually strand a single bag or lock anybody out of an area they needed. In at least two of those three zones, they reached the heart of the system and proved they could take it over. A working airport, mid-operation. Not a lab. That is how close to the crown jewels an attacker usually gets once they are inside.

And where do they land? The weakest, most forgotten machine on the floor.

> *HMIs are my favorite. I just take over your Windows XP box that cannot be patched, has no antivirus, and already connects to everything below it. Done. Impact created.* – A red team founder, on the show

Why a Firewall Is Not Enough

By now you might be wondering what the firewall was even doing this whole time. Good question. People treat the firewall like it is the wall, the thing that keeps the bad guys out. An OT network security expert I interviewed, who builds one-way gateway technology for the most sensitive networks on earth, gave me the best picture of what a firewall actually is.

> *A firewall is like a bouncer at a club. Show him a fake ID and he lets you in. Hand him a hundred-dollar bill and he lets you in. Shoot him and run right past.* – An OT network security expert, on the show

A bouncer is useful. But a bouncer can be fooled, bribed, or rushed, and here is the part that matters for anything critical. Once you are past him, that is often all you need. There is nothing behind him. And plenty of firewalls are not even doing the bouncer's job in the first place. An ICS security instructor on the show tests company after company that swears it has a secure perimeter, only to find the firewalls are passing everything right through.

> *If a firewall is not actually doing its job, you gained nothing. It is just a router. A really expensive router. A false sense of security.* – An ICS security instructor, on the show

How does that happen? Usually something broke, somebody added a rule that allows everything just to get the plant running again, and that temporary fix never got removed. So now the thing that says firewall on the box is quietly waving all the traffic through.

I am not exaggerating how routine that is. An OT threat researcher's team built their own industrial test lab, staffed by sharp people, and even there, every time something did not work, the automation engineer's first words were always the same. "It's the firewall." To keep the project moving, they would drop in a rule that allowed absolutely everything, just to prove the firewall was not the culprit. That is how it goes in the real world too, except nobody ever comes back to take the rule out. And an ICS security instructor I hosted has found firewalls so badly misconfigured that they did the attacker's job for him. One was set up to forward a single outside connection to all sixty-five thousand internal ports at once, so when he ran a routine scan, the firewall amplified it sixty-five thousand times and crashed itself into the ground. That was the device standing guard.

A firewall is necessary. It is one layer. It is not the wall, and it was never going to be enough on its own.

The Devices You Never Think About

The attack surface is bigger than your computers. It is the cable from the start of this chapter. It is the smart device on your shelf. One of my pentest guests once set up a digital picture frame for a family member, the kind you email photos to, and immediately started wondering whether he could sneak code in through a booby-trapped image and make the frame reach out and run it. It is a cheap, pocket-sized gadget you can buy online that can reach out and slam the brakes on a train. A government facility actually asked his team to try exactly that, and they could, because much of how trains signal and switch tracks is published in public technical documents for anyone to read. In one real incident overseas, pranksters with cheap radio gear broadcast the emergency stop command that a national rail system uses, the simple, unprotected signal every train is built to obey. Around twenty trains ground to a halt. Nobody hacked a computer. They just spoke the system's own language, and the system did what it was told. Cheap radio, real paralysis. And in the industrial world, it is the protocols themselves, many of which have no security at all. You can send a command to open a valve, and the equipment just does it, no questions asked, because it was built in an era when everything on the network was trusted.

I describe those unauthenticated commands like this. The machine hears a command and obeys, no questions asked, the stranger-in-your-kitchen problem I dig into in chapter eleven. A firewall watching that traffic sees a perfectly normal-looking instruction and lets it pass, because the instruction is normal. The problem is that nobody ever taught the machine to ask who is asking.

And Then They Send the Bill

They got in, they moved, and now comes the act, the part where they cash out. Most of the time that is ransomware, locking up your files and your systems until you pay. But it is getting more creative, and more physical. An OT network security expert I interviewed told me about attackers getting into building management systems, the computers that quietly run the elevators, the air conditioning, the sprinklers, and the fire alarms in a casino, a high-rise, or a university campus. The shakedown could not be simpler. Get into the fire-alarm system, and the message is, pay me fifty thousand dollars, or I trip the alarm and you evacuate this entire building, right now, during your busiest hour. And people pay, because emptying a packed casino costs a whole lot more than fifty grand. That is the act stage in the modern world. It is not always about stealing your data. Sometimes it is about holding the physical world hostage.

The Defender's Real Job

So if the attacker only has to be right once, and the firewall is just a bouncer, and the protocols obey anyone, what on earth do you do? You stop trying to be perfect, and you start being layered and watchful. You assume they will get a foothold eventually, because they might, and you build so that a foothold is not the end of the story. You segment your network so they cannot roam. You watch the traffic going out, not just coming in. And above all, you get fast at seeing them and shutting them down. An OT penetration tester I hosted framed the whole job perfectly.

> *They are only successful if we let them, if we are not monitoring and not responding. What has to speed up is the detection and the response.*
> *– An OT penetration tester, on the show*

Every technique in this chapter leaves a trace. The phishing click, the foothold, the sideways movement, the data walking out the door. If somebody is watching, you catch it while it is happening and you stop it. If nobody is watching, you find out from the ransom note. The difference between those two outcomes is not a bigger wall. It is paying attention.

One hard truth to hold onto while you build that detection. The old fallback, when something feels off I will just call the person and verify, is starting to crack. An OT network security expert I interviewed raised the uncomfortable version of it. When a voice can be cloned from a few seconds of audio, how do you actually know it is your coworker, or even your own mother, on the other end of the line? The answer is to verify on a channel the attacker does not control, and to lean on the family and team code words from the scams chapter. Verification still works. It just cannot rely on your ears alone anymore.

The Bottom Line

Real attacks are not magic. They are a process. Research, get in, move, act. They get in mostly by fooling a person or by sliding through a connection you forgot you left open, and they exploit holes that are often years old. Once inside, they move sideways using your own tools so they blend in, and they walk the data out a door you were not watching. The firewall you were counting on is a bouncer, not a wall, and sometimes it is just an expensive router. None of that is hopeless. Every step is visible if you are watching, and every step is harder if you have done the fundamentals. You cannot make yourself impossible to attack. You can absolutely make yourself hard to attack and quick to catch, and that is the whole ballgame.

Tools That Help

Categories that break the attack at each stage.

- **Phishing-resistant multifactor authentication.** Stops the stolen-password foothold cold, and a hardware key cannot be worn down by two in the morning push spam.
- **Secure remote access and network segmentation.** Close the forgotten doors, put every vendor connection under control, and box in any foothold so it cannot roam.
- **Patch and vulnerability management.** Because the holes being used are mostly old and already fixable. Close the known ones first.
- **Endpoint detection and response.** Tools that watch what is actually happening on your machines and flag the sideways movement a firewall never sees.
- **Email security and filtering.** Knock down the lures before they reach the tired human at the keyboard.
- **Network monitoring, including outbound traffic.** Watch the data trying to leave, not just the attacks trying to arrive.
- **Asset inventory.** You cannot watch, patch, or segment the device you forgot you owned.

Your Action Steps

1. **Turn on the strongest multifactor you can** for email, banking, and admin accounts. It defeats the most common foothold there is.

2. **Hunt down every remote connection into your world**, especially the vendor ones, and shut off or lock down the ones you cannot justify.
3. **Patch the old, known holes first.** Not the scary new headline. The five year old one nobody got to.
4. **Start watching the traffic leaving your network**, not just the traffic coming in. That is where your data walks out.
5. **Assume a foothold will happen, and plan to catch it.** Segment, monitor, and decide now who does what when something trips.

You do not have to be perfect. You have to be layered, and you have to be paying attention. Pick one stage of the attack and make it harder this week. Make the attacker work for every step, and most of the time they go looking for an easier door.

Part III

Securing Businesses and Organizations

Chapter Seven

Cybersecurity in the Workplace

Why Culture Beats Tools Every Time

One of my guests, a security leader I hosted who runs security for companies, told me about the day his director of IT walked up holding a question. Is that yours? Are you trying to get around the firewall? Somebody in the office had quietly mounted a wireless antenna, plugged it into the company network, and paired it with a cellular connection, building themselves a private little bypass around every security control, just so they could get to things faster without the firewall in the way. Wireless was flat-out banned in that environment. Somebody built their own anyway. And here is the thing he understood that a lot of security people never do. That employee was not a villain. He was just trying to do his job, and security had made the right way too painful, so he went and built a wrong way.

That is the whole story of cybersecurity in the workplace, in one antenna. You can buy every tool in the catalog, but if the people in your building experience your security as a wall between them and their work, they will go around it, and you can end up less safe than if you had done nothing at all. This chapter is about the part of security that has almost nothing to do with technology. The people, the culture, and the business they are all actually there to serve.

Make It Too Hard and They Route Around You

> *Make security too hard, and people will just find a way around it.* – *A security leader I hosted*

He would know, because he does it himself. Plenty of times in his career, when somebody cut off some access he needed, he just moved sideways through the system to get it back, because he was technical enough to. At one company he had a whole team who would come to him and say, hey, I need to get to such and such, and he would just go do it and let them sit at his computer. Not because he was reckless. Because the official path was broken and the work still had to happen. The more you tie people's hands, the more creative they get at slipping the rope.

A security consultant on the show, who has spent years on the security side, sees the same thing from a different angle. A team builds something brand new with no security baked in. When security catches it, the team says, relax, it is just a proof of concept, we will loop you in later. A month later, the proof of concept is in production. The business moves faster than your review cycle, every single time, and once the thing is live and working, nobody is going to rip it back out.

And the cost of friction is not only risk. This same security leader was once waiting for a meeting at a healthcare office and watched a receptionist book appointments. She had to scan her fingerprint over, and over, and over, about five minutes to book a single appointment, almost half an hour for a handful of them. Everyone there had simply accepted it as how the system works. Nobody saw that clunky security was quietly eating ten percent of that person's day. Good security does not just keep the bad guys out. It also gets out of the good guys' way.

You Are Not the Police. You Are the Plumbing.

Here is where a lot of security people go wrong. They show up as cops. He calls it the policeman stance. Here is the rule, we wrote it, the business does not understand. The trouble is, once you are the person who only ever says no, the business stops listening, and it does not matter how good your security advice is.

The reframe that changes everything is to treat security as a core business function, not a hall monitor. He has a great way of putting it. A hundred and fifty years ago, you could run a company with nobody doing accounting. It was possible. Then accounting became unthinkable to skip, because it turned out you needed somebody who could tell the profitable products from the loss leaders and spot the guy who was embezzling. Cybersecurity, he argues, is on that exact same path. It is not the thing the business exists for. It is the thing that keeps the business safe enough to do what it exists for.

I say it like this. We are the tail, not the dog. We support the business. We do not drive it. You cannot just lock the door and refuse to hand out keys, and then act surprised when people climb in the window. A CISO I interviewed, who has held the role twice, puts the same idea on the leader's shoulders.

> *You are not just a technology leader, you are a business leader, and you have to translate that into risk.* *– A CISO I interviewed*

Speak the Language of the Room

If security is a business function, then you have to talk like the business, and the most common way I see smart people fail is by talking right over everybody's head. A security consultant on the show told me about an aviation conference where a security expert from a big firm gave a talk on espionage to a room full of engineers and data analysts. Afterward, people came up to the consultant, not the speaker, and asked her, you are in cybersecurity, did you understand any of that? She admitted that even she had found it dense. Half the room had given up and gone to their phones, because the words never landed.

> *If you cannot explain it in the simplest terms for your audience, you might as well not say anything at all.* – A security consultant on the show

She wears a different hat for every audience. Executives want to know what you need, why you need it, and how much it costs. Developers want the one specific thing to fix, like the vulnerable library to update. Middle managers want resources and support. Same risk, three completely different conversations. It is not about buzzwords or showing everyone how smart you are. It is about being understood.

And the single most powerful translation there is, is dollars. I learned this asking for budget across more than forty plants. Humans are genuinely terrible at understanding probability. A one in fifteen chance means almost nothing to most people. But put a number on it. This twenty year old machine going down will cost us fifteen grand, the odds this year are this, so here is the expected cost of doing nothing. Now your CFO can actually make a decision, because you handed them something they understand. A CISO I interviewed boils the whole budget conversation down to one question.

> *Am I spending five dollars to protect one dollar?* – A CISO I interviewed

And when the number comes back bad, the same security consultant does something a lot of security people forget to do. She hands the decision back to the business. If there is a serious, exploitable gap, she does not just demand it be fixed. She asks the owner directly, are you willing to accept this risk? Most of the time, faced with owning that in writing, they say no, and now she has the support to go fix it. The risk was never hers to carry alone. It belongs to the business. Her job is to make it visible enough that the business can decide on purpose.

Decide What Risk You Are Willing to Live With

Here is a truth that makes purists uncomfortable. You are not going to fix everything, and you should not try. Some risks you accept on purpose, with your eyes open. The security leader I hosted sees this constantly with old machines that cannot be replaced, the twenty year old computer running a critical process, full of holes, that you cannot patch and cannot rip out. His answer is not to panic, and not to ignore it. It is to wrap it in what he calls a crunchy bubble. Do not let anybody plug a USB stick into it. Do not let it touch the internet. Watch it closely. And then accept the small bit of risk that remains, because you decided to, after doing the math.

What you cannot do is pretend. He describes two camps among the small manufacturers he works with. One camp connects everything and says nobody would bother coming after us, which is just hoping nothing happens. The other camp disconnects so completely that to move a file you physically carry it over on a drive, and yes, plenty of expensive machines still have a floppy disk slot for exactly that. He respects the second camp far more, not because cutting everything off is always right, but because they actually looked at their risk and made a choice. The danger is almost never the risk you accepted on purpose. It is the one you never looked at.

And do not confuse a clean audit with actual safety. Companies come to him a couple of weeks before an audit asking him to hand them a perfect compliance score. He tells them the truth. He can build a real program, but not in two weeks, and a perfect score you did not earn is not a shield, it is a liability. If you swore you had controls you did not have, and then you get breached, you have turned a bad day into a legal one. Compliance is a floor you have to clear, and why clearing it is nowhere near the same thing as being secure is the whole subject of chapter ten.

You Go Blind to Your Own House

A governance and risk professional I hosted does security assessments for a living, and his real superpower is seeing the things the people who work there stopped seeing years ago. The critical room with no camera pointed at it. The guest badge that somehow opened a door it should never have touched. The nice glass key cabinet with a tidy sign-out sheet, and the key to the cabinet left sitting right there in the lock.

> *When it is your own house, you walk past the same problem so often that you go blind to it.* – A governance and risk professional I hosted

I have my own version. I walked a brand new manufacturing plant, eight production lines, beautiful, and the network cabinets that ran everything were bolted fifteen feet up a pillar, behind tinted glass so you could not even see whether the lights were blinking, with no redundant switch, no redundant power, and no monitoring. If one failed, a whole line stopped. And right next to them hung a television, dangling by its own cord, because a forklift had clipped it at that same height. Nobody on that project had said a word about the cabinets, because everybody assumed somebody smarter had already thought it through. That is the trap. When you built the thing yourself and poured months into it, it is really easy not to see how ugly the baby is. The newest person in the room often sees the ugliest problem, and is the most afraid to point at it.

So build a culture where the junior person, the new hire, the intern, feels safe raising a hand and saying, have we thought about this? And when they do, thank them. The cost of an awkward question is nothing. The cost of the thing nobody mentioned can be the whole production line.

The same blindness lives in your paperwork. A governance and risk professional I hosted makes a sharp point about the incident response plan sitting on a shelf. It might be thirty beautiful pages that sail right through a compliance audit. But are the people named in it even still at the company? Has anyone actually read it? Because when everything is on fire, nobody is flipping through thirty pages to figure out what to do. A plan nobody has ever practiced is just a document. It looks like being prepared. It is not the same thing.

Your Vendors Are Your Risk Too

Your security perimeter is not your building. It includes every company you do business with. We already saw how a major retailer got breached through a heating and cooling contractor. And I have watched a crane inside the protected area of a nuclear plant get infected with malicious code, get cleaned, and then get reinfected eighteen months later, both times by a maintenance vendor's laptop walking in through the required scans, because a scanner can only catch the malware it already knows about.

So you have to manage your vendors like they are part of your own house, because for risk purposes they are. He spent years on both sides of this, sending the security questionnaires and answering them, the audits, the proof of controls. His philosophy is one I love. Do not hand a customer a fluffy document just to make them go away. Actually have the process, drink your own Kool-Aid, and then share it honestly.

And the risk is not only the outside vendor. Sometimes it is your own left hand not talking to your right. The same CISO told me about an automated process that nobody had bothered to walk through with the HR team. One day it fired off a termination for two hundred employees at once, and among the two hundred were the administrators of the very system that ran the terminations. He spent a holiday weekend putting it all back together. The fix was not a tool. It was a conversation he should have had first.

> *Sometimes you have to stop beating yourself up over it.* – *A CISO I interviewed*

One more shift worth knowing. Insurers will no longer write a cyber policy for a company that cannot prove the basics, a change we will come back to in the next chapter.

Two more workplace realities deserve a straight sentence each. First, your office is not a building anymore. Half your people work from kitchen tables, on home routers you have never seen, sometimes on the same network as the kids' game console, so your culture and your controls have to travel home with them. The same basics from the start of this book, applied to every laptop that leaves the building. Second, not every insider is well-meaning. The vast majority are, and I will keep saying so. But the disgruntled employee walking out the door with a grudge and a working login is real, and it is exactly why access gets shut off the day somebody leaves, not the week after, and why nobody holds more access than their job needs while they are there.

Culture Eats Tools for Breakfast

All of this rolls up into one idea that is hard to sell and impossible to skip. You cannot buy your way to secure. The same security consultant says it as plainly as anyone.

> *You can own the most advanced technology in the world, but without a culture of security, technology alone will not protect your company.* – A security consultant on the show

The best comparison I have heard for this came from a CISO on the show, and I have used it ever since. Look at modern medicine. We have a specialist for every organ, but who is looking at the whole body? Sometimes you do not need a heart transplant. You need somebody who can see the whole picture and tell you to cut back on the salt. Security spent years splitting into narrow specialists, the firewall person, the identity person, the antivirus person, and somewhere in there we lost the people who understand how it all connects. The best security cultures grow generalists who can see the whole body, not just deeper and deeper experts each guarding one fence post.

He calls it a team sport, and he learned the hard way that you cannot play it alone or play it as a jerk. Early in his career, by his own account, he was the smartest, most abrasive guy in the room, hammering his point into everyone's head. At his first job as a security leader, the company ran a 360 review, where your peers and your team get to tell you honestly how you actually come across. He says you find out exactly how much of a jerk you are, if the people answering are honest. The turning point was getting his ego out of the way long enough to hear it. That is what turned him from a smart technician into a leader people would actually follow.

And culture, or the lack of it, shows up in the most boring place imaginable. I have walked into companies sitting on millions of dollars of security tools that were never even turned on.

> *Far too many companies own millions of dollars of software and hardware that is sitting on a shelf, never even switched on.* – A CISO I interviewed

The tool was never the answer. It was always the people who would, or would not, actually use it.

The Bottom Line

The hardest part of workplace security is not the technology. It is the people, and the culture you build around them. Make security too hard, and your own employees will tunnel around it. Show up as the police, and the business will tune you out. Talk over everyone's head, and nothing gets funded. The programs that actually work treat security as a core business function, translate risk into the dollars leaders can act on, stay humble enough to see their own blind spots, manage their vendors like family, and build a culture where security is everybody's job instead of the department of no. Buy the tools, sure. But the tools were never going to save you. The people were.

Tools That Help

The categories that build a security culture, not just a security stack.

- **Security awareness training.** Ongoing and real, not a once-a-year video nobody watches. Your people are the most targeted layer, so make them a trained one.
- **Identity and access management with least privilege.** Give people exactly the access they need, no more, so the painful workarounds stop being necessary.
- **Single sign-on and usable security.** Make the secure path the easy path, so doing the right thing is also the convenient thing.
- **A governance, risk, and compliance process.** So risk decisions get made on purpose, by the business, and written down, instead of by accident in the shadows.
- **Third-party and vendor risk management.** Know exactly who has access to your world, and hold them to a real standard.
- **Tabletop exercises.** Practice as a team, so the first time you respond to a crisis is not during the actual crisis.
- **Cyber insurance, backed by real controls.** A backstop for the worst day, never a substitute for doing the work.

Your Action Steps

1. **Find where your own people are routing around security**, and fix the friction, not the person. That workaround is telling you something.
2. **Put a dollar figure on your top three risks** and take it to leadership in their language, not yours.
3. **Walk your own environment like an outsider**, or bring one in, and write down every ugly baby you have gone blind to.
4. **List every vendor with access to your systems** and hold each of them to a standard you can verify.

5. **Empower your newest people to raise a hand**, and thank them out loud when they do.

Security is not a wall you build once. It is a culture you tend every day, and it will guard the company when you are not looking. Pick one of these and start this week.

Chapter Eight
Ransomware

The Day Everything Locks Up

It was just after midnight, and a security leader at a 911 dispatch center I interviewed was getting ready for bed when his phone started buzzing. Servers were going offline. At first he was not too worried, because servers reboot and life goes on. But these were not coming back, and there was no pattern to it, random machines spread across different systems, which ruled out all the easy explanations. He connected in from home and went to work, and as he watched one server's health on his screen, it blue-screened right in front of him. That is the moment his stomach dropped. The way he described what went through his head stuck with me. He was the cybersecurity guy, and he had finally gotten ransomware. He was finally the victim.

He had not. What that security leader was watching that night was the big global software outage of 2024, a faulty update that crashed millions of computers around the world in a matter of hours, with no attacker anywhere in sight. But I love that his mind went straight to ransomware, because it tells you exactly how that word sits in the gut of everybody who runs a network now. It is the thing we are all quietly waiting for. So let me tell you what it actually is, what it actually does to you, and the surprising truth about how you beat it.

What Ransomware Actually Does

The mechanics are simple and brutal. Malware gets onto your systems, usually through one of the doors we covered a couple of chapters ago, a phished employee, a stolen password, a forgotten remote connection. Then it quietly spreads, and when it is ready, it encrypts everything, your files, your servers, your backups if it can reach them, and locks you out. Then comes the note. Pay us, usually in cryptocurrency, and maybe we give you the key. More and more, they also steal a copy of your data first and threaten to leak it if you do not pay, so you are squeezed from both sides.

And remember, this is not a lone genius. It is an industry, a service you can rent, which is why it has been the number one threat for years on end. A CISO I interviewed put the monotony of it perfectly.

> *I gave a keynote six years ago and the top threat was ransomware. Last year, ransomware. The latest breach report, still ransomware.* – *A CISO I interviewed*

Here is the part people get wrong, though. The ransom is not the expensive part. The downtime is. That is the number I keep hearing from the people who work these incidents. The real cost is driven by the production outage, how long you were down and how much business you lost while you were. The check to the criminals is often the cheapest line on the whole bill.

When Pulling the Plug Is Not an Option

In a normal office, the standard ransomware advice is isolate and wipe. Pull the infected machine off the network and rebuild it. But in the physical world, where computers run turbines and pumps and furnaces, that same advice can get somebody hurt. An OT security consultant I interviewed walks into industrial sites with ransomware spread across the floor, and the operators ask him the obvious question. Do we pull the plug? His answer is another question. What happens if you do?

Because that infected machine might be the one screen an operator is using to control a live steam turbine. Yank it, and now nobody can see or control a giant spinning machine full of energy. The cure becomes more dangerous than the disease.

> *Maybe if I pull this screen off, I lose control of my steam turbine. Maybe we should not do that.* – *An OT security consultant I interviewed*

That is why, in the industrial world, you find old malware just living in the walls. The OT security consultant turned on monitoring at a plant and watched it light up with WannaCry, the infamous worm, on thirty and forty machines at once. The customer's response? We are still running production. When we get to it, we will get to it. One of his colleagues found a plant living with an old Conficker infection like it was bed bugs, just something you coexist with, because taking the system down to clean it would stop the money.

> *In IT you live with zero days. In OT you live with forever days.* – *An OT security consultant on the show*

And here is the cruelest twist. You do not even have to be the one who gets breached. A very common pattern is that ransomware hits a company's office network, never touches the plant at all, and the company shuts the plant down anyway, because they cannot be sure it is safe. The attacker hit the business side and got a production outage for free. The company can tell the world, truthfully, that their operations were never breached, and still be completely shut down. Same outcome, either way.

There is a whole world of computers running turbines and pumps and furnaces like that, and we are going to spend three chapters inside it starting in chapter eleven.

To Pay or Not to Pay

Eventually the hard question lands on somebody's desk. Do we pay? I am not a lawyer, and I am not going to tell you what to do, but I will tell you what belongs in the decision. Paying does not guarantee you get your data back, because you are trusting a criminal to keep their word. It directly funds the next attack, on the next victim. And if the group behind it is on a government sanctions list, paying them can itself be illegal, and now you have two problems. Plenty of organizations pay anyway, because every hour of downtime is costing them more than the ransom, which is exactly the trap the attacker engineered. The only real way to win this decision is to never have to make it. And that comes down to one deeply unglamorous thing. Recovery.

And whatever you decide, do not decide it alone. The moment you know you have been hit, three calls belong at the top of the list. Call law enforcement, because the FBI has seen your attacker before and sometimes has decryption keys or intelligence you do not. Call a lawyer who handles breaches before you touch, wipe, or rebuild anything, so you do not destroy evidence or step on a legal landmine while you are trying to help. And call your insurer, because a lot of cyber policies require you to notify them and get their consent before you pay a ransom or hire a recovery firm, and skipping that call can turn a covered claim into a denied one. Put those three numbers in your incident plan today, on paper, so you are not hunting for them on the worst morning of your year.

A quick word on the safety net people reach for here. Cyber insurance. It helps, but do not mistake it for a rescue. A guest I interviewed who works in cyber insurance describes it as the cherry on top, not the meal. A policy might pay back ten or twenty percent of what an incident actually costs you, once you count the downtime, the lost business, the cleanup, and the hit to your reputation. It is real money and worth having. But it is not going to make you whole, and it is definitely not a substitute for being able to recover on your own.

The Real Answer Is Resilience, Not a Perfect Wall

Here is the mindset shift that changes everything about ransomware. Stop trying to be unhackable, because you will not be. Start trying to be recoverable. The same security leader, who runs technology for a 911 dispatch center, draws the line as cleanly as anyone I have heard.

> *Reliability is protecting against failure. Resiliency is your ability to recover from failure and get back into operation.* – *A security leader at a 911 dispatch center*

So come back to that security leader's bad night. Yes, that outage took down a live 911 center, the place people call when it is the worst moment of their lives. But he had built for resilience. His alerting caught it fast, he found the cause, and he and his team worked out their own temporary fix before the vendor's guidance reached them. Total time to recover, by his own count, just over an hour, sixty-two minutes and change. And during those sixty-two minutes, his dispatchers fell back to the contingency plan they had actually practiced. Pen and paper. They wrote the active calls down by hand, dispatched them over the radio, handled real emergencies, and the fire departments they serve never knew a thing was wrong.

> *Our contingency plan was pen and paper. The fire departments had no idea. We never missed a 911 call.* – *A security leader at a 911 dispatch center*

That is resilience. Not a wall that never breaks, but a plan for the day it does. And the after-action lesson is pure gold. He discovered that his alerting server was running on the very system the outage killed, so the moment things went down, his alerting went down with them. Now he keeps alerting somewhere completely separate. You only learn that by living through it, and to his credit, he wrote the whole thing up and shared it publicly so other agencies would not have to learn it the hard way.

The flip side of that dispatch center leader is the company that never planned at all. My cyber insurance guest told me about a regional grocery chain headquartered near him that, after a hurricane put its data center under several feet of water, could not take a credit or debit card for almost two weeks, because it had no backup way to process payments. Aisles full of food, customers ready to buy, and no way to complete a sale, for two weeks. They had simply never imagined a flood taking out that one system. That is what the absence of resilience looks like. The dispatch center leader, by contrast, rode out a hurricane that left his whole region without power for four days, running on his own generators and battery backup, and his dispatchers barely noticed. Same storm season, two completely different outcomes, decided long before the storm by whether somebody had planned for the bad day.

Backups You Have Not Tested Are Not Backups

The spine of all recovery is backups, and there is one sentence in this whole chapter I want you to tattoo somewhere you will see it.

> *If you are running backups but never testing them, you do not actually have backups.* – *A security leader at a 911 dispatch center*

Everybody backs up. Almost nobody tests whether they can actually restore. And ransomware is precisely the day you find out the hard way that your backup was corrupt, or incomplete, or that it got encrypted right along with everything else. An incident response leader on the show, who commands incident response for a living, warns about the trap waiting on the far side of a sloppy recovery. You restore from a backup you never checked, and you reintroduce the very malware you were trying to get rid of, and now you are right back at square one, doing the whole painful recovery a second time. Nobody wants to do that twice.

So the rules are simple, even if doing them is not. Keep more than one copy. Keep at least one copy offline, physically disconnected, where ransomware cannot reach across the network and encrypt it too. And actually practice restoring, on a normal Tuesday, before you are desperate. A backup you have never once restored is a hope, not a plan.

Resilience also means making deliberate tradeoffs ahead of time. When that security leader built his 911 center, he had to decide whether to encrypt the hard drives on the dispatch workstations, which is great if a machine gets stolen but slows down recovery. He looked at the real threat, those machines sit in a locked, guarded facility, decided the theft risk was genuinely low, and chose not to. When the outage hit, that single decision meant a clean, fast recovery instead of fighting encryption on every machine. The point is not that encryption is bad. It is that he chose on purpose, with recovery in mind, instead of just flipping on every setting and hoping for the best.

The Update That Was Not an Attack

Let me come back to that 2024 outage one more time, because it taught a lesson that applies straight to ransomware. It was not an attack. It was one vendor's update, pushed automatically to millions of machines at once, that broke them all at the same instant. The same cyber insurance guest watched companies scramble, and the problem was not that the fix was hard. It was that they had to physically walk up to every single machine to apply it. The cry for help was the same everywhere. I need bodies. We know how to fix it. I just do not have enough people to get to every machine in time.

Two lessons fall right out of that. One, depending on a single vendor for everything means that vendor can quietly become your single point of failure. Two, control your own updates. The nuclear and industrial folks I called that day who ran the very same software were fine, because they did not auto-update. They tested first. As an asset owner, my rule was simple and a little old-fashioned. We do not patch unless we are sitting right in front of the machine, and we never, ever patch on a Friday. Because if a remote update breaks a plant that is three hours away, that plant is down until somebody gets in a truck and drives there.

And do not skip the boring maintenance to save a few dollars, because that bill always comes due with interest. There is a line for the rhythm of it that I heard on the show and never forgot. If Patch Tuesday does not get them, Exploit Wednesday does, meaning the moment a fix gets announced, attackers know exactly which hole to go hunting for in everybody who has not applied it yet. Skipping your updates and your basic hygiene to save money is like never changing the oil in your car to save money. It works right up until the engine seizes, and then you are not paying for an oil change. You are paying for an engine.

When It Happens: Respond Like a Team, Not a Mob

Someday you will have an incident, and how you handle it is mostly about people, not tools. The same incident response leader has a first rule that surprises people. It is blameless. You do not go hunting for the person to fire. You fix the process that let it happen.

Here is why that matters so much. The incident response leader told me about the old days of the data center, when an engineer thought he was logged into one server, was actually on another, and rebooted the wrong box, taking down live production. And then he did not tell anybody. First because he did not realize, and then because he hoped he could quietly fix it before anyone noticed. There was no culture of safe reporting, so the problem hid, and hiding made it worse. The fix was not punishment. It was process. They put the machine's name right on the screen, and they adopted the three-way communication that the military and plant operators use. I heard you say reboot server one. Yes, reboot server one. Okay, rebooting server one. Boring, and it saves you.

> *I try to run a blameless post-mortem. Blaming the team or the person is never constructive.* *– An incident response leader on the show*

So when somebody clicks a bad link and comes to report it, the incident response leader's team says thank you, first thing, before anything else. Because fast reporting shrinks the blast radius, and a person who is too scared to report is the most dangerous thing in your building. During the chaos itself, he assigns a scribe, somebody whose only job is to write down every we-should-fix-this so the lessons are not lost, and he treats those lessons like money, tracked and actually spent. And he practices the whole thing with tabletop exercises before the real fire, for the same reason we make kids walk through fire drills. So that the first time is not the real time.

> *Nobody gets to breathe until things are contained.* *– An incident response leader on the show*

The Bottom Line

Ransomware is the nightmare everybody is bracing for, and it is not going away, because it is a business, and business is good. But the headline number, the ransom itself, is the part that matters least. The real damage is the downtime, and the real defense is not a perfect, unbreakable wall. It is resilience. Assume you will get hit. Then make sure you can recover, with backups you have actually tested, an offline copy that ransomware cannot reach, an incident plan you have rehearsed, controlled updates that cannot break everything at once, and a blameless culture where people report fast instead of hiding. Do that, and the worst day of your year becomes a sixty-two-minute story you tell at a conference, instead of the day your business does not come back.

Tools That Help

The categories that get you from victim to recovered.

- **Tested, offline, immutable backups.** The foundation of all of it. Multiple copies, at least one disconnected, and proven by an actual restore.
- **Endpoint detection and response.** To catch the encryption starting and the attacker moving, before the whole place goes dark.
- **Network segmentation.** So one infected machine cannot become every machine. It is the difference between an incident and a catastrophe.
- **Multifactor authentication.** To shut down the stolen-password foothold that kicks off most of these attacks in the first place.
- **An incident response plan, and ideally a retainer.** Know who does what, and have expert help on speed dial before you need it, not after.
- **Tabletop exercises.** Rehearse the bad morning as a team, with the people who would actually be in the room.
- **Out-of-band communication and alerting.** So when the systems go down, you can still coordinate, and your alerting does not die along with everything else.

Your Action Steps

1. **Test a restore from your backups this month.** Do not assume it works. Prove it works.
2. **Make sure one backup copy is offline**, physically disconnected, where ransomware cannot reach it.
3. **Take control of your updates.** Test before you deploy, and do not push to production on a Friday afternoon.
4. **Write a one-page incident plan.** Who to call, what to do first, and how you communicate if the systems themselves are down.
5. **Run one tabletop.** Walk through a ransomware morning, out loud, as a team, before you ever have to live it.

You will not stop every attack. You can absolutely decide whether the next one is a catastrophe or a footnote. Pick one of these and do it this week.

Chapter Nine
Securing Small Businesses

Too Small to Matter Is a Myth

I hear it all the time. I am too small. Nobody would bother coming after me. I have nothing they want. I understand why people believe it, but it is one of the most dangerous things a small business owner can think, because it is exactly backwards. The attackers are not skipping you because you are small. A lot of the time they are coming for you precisely because you are small, since small usually means you do not have the budget, the staff, or the know-how to put up much of a fight.

Here is the part people miss about how this actually works. The bad guys are not sitting around hand-picking targets worthy of their attention. They are running automated tools that scan the entire internet looking for anything easy, the way a fisherman casts bait and waits for whatever bites. That is why we call it phishing. They are not choosing which fish to catch. They throw the net, and if you are the little fish that bites, you are just as caught as the big one. Bad actors do not care who you are, what you are, or how big you are. They are coming after money.

And the numbers are not on your side. By various estimates, something like four in ten cyberattacks target small businesses, not the giants. And here is the part that should make you put this book down and go change a password right now. For a lot of small businesses, a serious breach is an extinction-level event. Many that get hit hard, hit where it really counts, never reopen. Not slowed down. Gone.

I once assessed a manufacturing company in its very first year of operation. Good size, hundreds of employees, a real factory, millions in revenue, and actually profitable. By any measure, a success story. But I looked at their situation and realized that if they had a cyber event and lost even a few days of production, maybe a week, they would close their doors. They simply could not absorb it, not that early in their growth. A big company can take a punch like that and limp on. For a young small business, a few hours of downtime can be the whole ballgame.

And it is not just established companies. There was a wave of articles a while back about startups skipping cybersecurity to save money in those scrappy early days, and it makes me wince every time, because it can be the very thing that kills the company. You can have a brilliant product and a great idea, but take one ransomware hit you cannot recover from and the idea goes out the window along with the business. The hard truth I tell every owner is the same. It is not a matter of if a cyberattack is coming. It is a matter of when, and the only questions that matter are how big the impact is and how fast you can recover.

The Security Poverty Line

There is a phrase for the companies caught in this trap, and I first heard it framed this way by a cyber risk advisor I hosted, someone who does cybersecurity for the construction world. He talks about the security poverty line. Above it are the companies with the budget and the people to defend themselves. Below it are the ones who cannot afford a security team, cannot afford the fancy tools, and often do not even have a real IT person. He calls them cyber orphans, and he says they are everywhere.

That same advisor told me about the CEO of a sizable electrical contractor who said to him, flat out, if there is a cyber scam out there, we have fallen for it. We have been hit by everything. And these are not careless people. They are builders, busy running a real business, who never once signed up to also be security experts.

That is the other half of the small business reality. He describes the typical firm. The owner's dad or granddad started it, they are doing tens of millions in revenue with a couple hundred employees, and they would honestly rather run the whole thing on a pencil and a piece of paper. When computers showed up, they tapped the youngest person in the office to be the computer guy, because you are young, you know this stuff. That young person is now quietly responsible for the security of the entire company, with no training and no budget, and is often terrified to admit it.

Borrow the Safety Sign

Here is something I love about how this same cyber risk advisor approaches the construction world, because it works for any small business. Walk onto any job site and there is a sign by the entrance counting the days since the last safety incident. Everybody sees it. Everybody is a little proud of the number. That safety mindset is baked right into the culture. His whole pitch is to borrow that exact instinct for cyber. Put up the equivalent board. Make not getting hit something the team notices and takes pride in, the same way they already do with hard hats and steel-toed boots.

The trick is meeting people where they actually are. He shows up to these firms in boots and jeans, and they call him the high tech redneck, because the last thing a builder wants is a lecture from somebody in a pocket protector who has never set foot on a site. You cannot scold a small business into being secure. You have to make it part of how they already work and how they already talk.

And do not let anyone tell you their old-school business does not really run on technology. He asks the crane operators who push back on all this newfangled stuff a simple question. Do you have any idea how much technology and data are packed inside that crane you climb into every day? Just the crane. The pencil-and-paper company does not really exist anymore. Everybody is a technology company now, whether they meant to be one or not, which means everybody has something worth protecting.

Nobody Gets Hacked Alone

Here is the part that should change how a small business owner thinks about all of this. You are not just a target because of what is in your own bank account. You are a target because of who you do business with. That cyber risk advisor put it in a way I love.

> *Nobody gets hacked alone.* – *A cyber risk advisor, on the show*

Remember the retailer from chapter four, breached through its little heating and cooling contractor. The little guy was the way in. So if you are a small business that supplies, services, or connects to a bigger one, you are not too small to matter. You are the soft spot they are counting on.

And the path can be heartbreakingly simple. A governance and risk advisor on the show, who works with logistics and supply-chain companies, describes it plainly. Somebody in accounting opens a bad email, the firewall was never set up right, and there is a clear path from the office systems straight into the part of the business that actually moves the goods. One click, one misconfiguration, and the whole thing is connected. The good news is that the bigger companies, and the insurers, are starting to push back. Surety bonds and cyber insurance now want proof that you have real controls, so the pressure to get this right is coming for you whether you welcome it or not.

Two flavors of this deserve their own warning for a small business. The first is the fake-invoice game, the corporate con we met back in the scams chapter, business email compromise. Somebody impersonates your boss or a vendor by email and quietly redirects a payment, and it quietly costs businesses more money every year than the ransomware headlines do. The fix is a rule, not a tool. Any change to payment details gets verified by phone, on a number you already had. The second one stings a little. The very IT provider you hired to keep you safe can be the way in, because attackers love managed service providers for the same reason you do. One login reaches a hundred businesses at once. So ask yours how they secure their own access to you. If the question offends them, that is an answer too.

Good Enough, and Just Start

So if you are below that line, where do you even begin? The mistake I see is people staring at the giant mountain of everything they should do, getting overwhelmed, and doing nothing at all. I think of it like getting my freshman English paper handed back, covered in so much red ink that I had no idea where to even start. If you overwhelm somebody, they freeze.

That cyber risk advisor has exactly the right mindset here. You are not trying to become an impenetrable fortress. You are trying to go from a three out of ten to a five or a six. Fix the obvious stuff. If everybody in the company has full administrator rights on their computer, take that away. Make it a little harder for the attacker. That is it. That is the whole first step.

> *Security is a game of good enough. Just find a place to start.* – A cyber risk advisor, on the show

And a quick word on the word free. There are great free and open-source security tools out there, and I use plenty of them. But free is not actually free. Somebody still has to set it up, run it, and keep it running, and for a small business with no spare hands, sometimes paying for a tool that just works, with real support behind it, is the cheaper choice once you honestly count your own time. Be realistic about what you can actually maintain, not just what you can download.

The Cheap Stuff That Actually Works

Here is the genuinely good news. The handful of things that stop the most attacks are cheap, and some of them are free. You do not need a big budget to be a hard target. You need to actually do the basics, the same ones we keep circling back to in this book.

- **Turn on multifactor authentication everywhere.** It is usually free, and it shuts down the stolen-password attack that kicks off most breaches.
- **Use a password manager.** And here is a bonus most people miss. It is also your business continuity plan. If your one IT person gets hit by a bus, or wins the lottery and walks out the door, you are not digging through sticky notes praying you find the passwords. They are all in one safe place.
- **Take away admin rights.** Most people do not need to install software on their work machine. Removing that one privilege stops a huge amount of malware cold.
- **Segment your network, even on cheap gear.** I do it in my own house, one network for work, one for the kids, one for the smart devices, one for guests, so a hacked smart gadget cannot reach the laptop with the company books on it. A small shop should split its point-of-sale, its office computers, and its guest Wi-Fi the very same way.
- **Filter your web traffic.** A content-filtering service blocks the known-bad sites for the whole company, often for next to nothing.
- **Back up, the right way.** The three-two-one rule. Three copies of your data, on two kinds of media, with at least one of them off-site. For a really small business, that can be as simple as backing up to a drive, putting the drive in a fireproof safe, and keeping it somewhere else, so a fire, a flood, or ransomware cannot take your data and your backup in one shot. The worst version of this is the company whose only backup lived on the very server that just died.
- **Train your people.** The cheapest control there is. Your employees are the most attacked part of your company. Ten honest minutes about how scams actually work is worth more than most software.

And you do not need a fancy system to stay organized. The governance and risk advisor suggests something as simple as a one-page list of what needs fixing or patching, kept current, so the thing an engineer flagged six months ago does not quietly get forgotten. Cheap, boring, and effective. That is the whole theme.

Talk About It Like a Business

The last shift is the most important, and it costs nothing. Stop thinking about cyber as a technology problem and start treating it as a business risk, right alongside the financial, legal, and reputational risks you already manage every single day. The governance and risk advisor put it well.

> *You have to talk about cyber in business terms. Financial, reputational, legal. Cyber touches all of them. You just adapt the language.* – *A governance and risk advisor, on the show*

Because in the end, this is a math problem like any other. I would never tell a small business to spend more on security than the thing they are protecting is worth. If I spend more than I make trying to be secure, I might as well shut the doors myself. The goal was never zero risk. There is no such thing as zero risk.

> *Nobody can eliminate all risk, but you can manage it.* – *A governance and risk advisor, on the show*

And managing risk includes deciding, out loud and on the record, which risks you are choosing to live with. The governance and risk advisor makes a great point that even the paperwork has value here. When the owner or the board looks at a risk and consciously decides to accept it, and somebody actually writes that down, you have done something real. You may not love the risk, but it was a decision made on purpose, with eyes open, instead of a surprise that blindsides you later and leaves everyone asking how nobody saw it coming.

And manage is the key word in that sentence. You will not stop everything. So put your energy where it actually counts, into being able to take a hit and keep going, or get back up fast. For a small business, resilience is not a luxury item. It is survival.

Plan for the Bad Day Before It Comes

Whatever your budget, the cheapest thing you can possibly do is think about the bad day before it arrives. There is a phrase I borrowed from the military world for it, planning left of bang. If bang is the moment everything goes wrong, then everything to the left of it is preparation, and everything to the right is reaction. The more you do on the left side, the less of a scramble the right side becomes. When an attack finally hits, you want it to feel like, okay, we thought about this, here is what we do. Not, oh no, what now.

Nobody plans perfectly, of course. As the saying goes, everybody has a plan until they get punched in the face. But having any plan at all still tells you which way to run when the punch lands. A small business with a one-page we-got-hit checklist will be recovering while the one figuring it out from scratch is still standing in the parking lot panicking.

And part of planning is knowing what you actually depend on, because it is almost always more than you think. The governance and risk advisor told me about a financial firm in Bermuda that got knocked completely offline on September eleventh, two thousand one. Not by the attack itself. A key undersea cable that carried their connection happened to run through the affected area and got cut, and just like that they went dark, for a reason nobody in that company had ever imagined. We all lean on infrastructure we never see and never think about. The cables, the cloud, the one vendor, the one server. Resilience starts with being honest about those hidden dependencies, and having even a rough plan for the day one of them quietly disappears.

The Bottom Line

You are not too small to be attacked. You are very often exactly the right size, because small so often means easy, and easy is precisely what the automated nets are trawling for. But you are not helpless, and you do not need a big budget. The handful of cheap things, multifactor authentication, a password manager, network segmentation, real backups, and a trained team, will lift you above the line where most attacks shrug and move on to somebody else. Do not try to do everything. Just start, fix the obvious, and treat it like the business risk it is. Because for a small business, the difference between a rough week and a closed sign can come down to whether you did the basics before the bad day, or wished you had after it.

Tools That Help

Categories, cheap to start, that punch above their weight.

- **Multifactor authentication.** Free, and the single highest-impact thing on this list.
- **A password manager.** Security and business continuity in one tool.

- **Tested backups, with one off-site or offline copy.** The three-two-one rule, proven by an actual restore.
- **DNS and content filtering.** Block the known-bad cheaply, for the whole company at once.
- **Network segmentation.** Even on consumer gear, to keep one bad device from reaching everything you have.
- **Endpoint protection, or a managed security service.** If you cannot run it yourself, rent the expertise. That is a perfectly good answer.
- **Cyber insurance and basic vendor vetting.** A backstop for the worst day, and a way to check who you are connecting to.

Your Action Steps

1. **Drop the too-small myth.** Accept that the automated attacks are coming for whatever is easy, and decide not to be easy.
2. **Do the three cheapest wins this week:** turn on multifactor authentication, roll out a password manager, and take away admin rights.
3. **Set up real backups** with one copy you can disconnect, and prove you can actually restore from them.
4. **Segment your network**, even with the equipment you already own.
5. **Have one honest conversation with your team** about how scams really work. It is the best money you will never spend.

Small does not have to mean defenseless. Do the few things that matter and make them go bother someone else. Pick one of these and do it this week, before you need it.

And that pressure I mentioned, the insurers and the big customers demanding proof, is really just the small version of something much bigger. Governments noticed too, and that is where we go next.

Chapter Ten

Governments, Regulations, and Why Compliance Is Not Security

Passing the Audit Is Not the Same as Being Safe

A power generation cybersecurity leader I interviewed spent years walking into power plants and asking a simple question. Do you have any remote access? And almost every time, the plant would say, nope, none, we are isolated out here. Then he would look around, and over in the corner he would spot a bank of dial-up modems quietly humming away. Oh, those? he would ask. Those are just for our third-party vendor, so they can dial in. From where? From Georgia. So, he would say, a vendor in Georgia can dial directly into your plant. That is remote access. The plant was not lying to him. They genuinely did not think of it that way. The word meant something different to them than it did to him.

That little gap, between what a rule says, what people think it says, and what is actually true on the floor, is the whole subject of this chapter. Governments and regulators have stepped in to force organizations to take security seriously, and that is mostly a good thing. But there is a trap sitting right in the middle of it, and it is the single most important idea in this chapter. Passing the audit is not the same as being safe.

Compliance Is a Floor, Not a Ceiling

Let me say the thing that gets security people in trouble at conferences. Compliance does not equal security. You can be one hundred percent compliant, pass every audit with a gold star, and still be wide open, because the things a rule requires are never the same as everything you actually need to do.

I have said for years that ninety-nine point nine nine nine percent compliant is not compliant, and one hundred percent compliant is not necessarily secure. Those are two completely different yardsticks. One measures whether you did the specific things on the list. The other measures whether you are actually hard to attack. They overlap, but they are not the same, and confusing the two is exactly how an organization talks itself into a false sense of safety.

Part of the problem is that the rules are written deliberately broad, so they can apply to a huge range of organizations. Which means two companies can read the exact same requirement and do completely different things, and both pass. I compare it to religion. Two people read the same line in the same book and walk away with two completely different interpretations, each one certain they have it right. Compliance language is just like that. Same words, very different plants.

And a lot of these rules are, by design, you grading your own homework. One guest, a vulnerability management researcher on the show, makes the point that a standard like the grid rules largely comes down to doing what you said you were going to do. You write down your own plan, and then you get audited against your own plan. That can work beautifully for a mature organization that writes an honest, ambitious plan. But it also means an organization can write itself an easy plan, meet it perfectly, and pass with flying colors, while remaining wide open. The audit checks that you did what you promised. It does not check whether what you promised was anywhere near enough.

Scope Is the Loophole

Here is where the gap turns into a real, gaping hole. Most regulations only cover part of your environment, the assets that fall inside the scope of the rule. And human nature does the rest. The same power plant security leader who found those dial-up modems put it plainly. If a company has ten assets and only seven are in scope for the regulation, they secure those seven, do the bare minimum to pass on them, and leave the other three out in the cold. And then they tell themselves, I did my compliance, so I am good.

But the attackers do not respect your scope. Remember the big retailer breached through a heating and cooling vendor, or the pipeline shut down by ransomware that hit the business side, not the controls. In both cases, the thing that got them was sitting outside the part everybody was busy watching. A chain is only as strong as its weakest link, and the weakest link is almost always the thing that was conveniently not in scope.

That same power generation cybersecurity leader has the perfect mantra for hunting down those gaps. He spent years filling out security surveys that asked things like, do you use antivirus? And the honest answer was always, yes, except for where we do not. That little exception is exactly where the risk lives.

> *I am trying to uncover that we are good, except for where we are not. Because I cannot fix problems we do not know about.* – A power generation cybersecurity leader, on the show

He learned to measure instead of assume. He would have sworn his company used multifactor authentication for every single remote user, until they actually measured it and found four users who never had it turned on. Four open doors nobody knew about, in a place everybody believed was fully covered. You do not find those by reading the policy. You only find them by checking the reality.

The Audit Loves a Checkbox

There is a deeper reason compliance and security drift apart, and it comes down to what an auditor can easily verify. A vulnerability management researcher on the show, who builds vulnerability tools for utilities, explains it beautifully. Applying a patch is binary. You either installed it or you did not, and an auditor can check a box. But often the smarter, safer move in an industrial environment is not to patch at all. It is to mitigate, to put a firewall rule in place or turn off the risky service, so the vulnerability simply cannot be reached. That is genuinely better security. But it is muddy. It is hard to prove on a checklist. So the rules, and the people enforcing them, drift toward the thing that is easy to audit, the patch, even when the better answer is the one that is hard to check.

That is the box-checking trap in a nutshell. And it is why one guest, who helped write one of the big power-grid security standards, has a rule he repeats to his own teams. Real success is not a perfect score on a shrunken test. It is knowing the true status of everything, including the ugly parts you would rather not look at.

> *Success is not just accomplishing something. Success is knowing what the status really is. Do not shrink the scope just to claim a hundred percent.* – A power generation cybersecurity leader, on the show

Regulation Can Be Built Well or Badly

Now, I do not want to leave you thinking regulation is the enemy. It is not. Done right, it is one of the most powerful forces for good in this entire field. The trick is all in how it is designed, and I have a front-row story for both versions.

Here is the part that gives him the right to say all this. He actually sat on the team that wrote one of the big power-grid security standards, the rules known as NERC CIP. In an early version, an asset was either high-impact, with a mountain of requirements piled on it, or it was nothing at all, with none. Guess what that produced. Companies poured enormous money and effort into protecting the one or two assets that crossed the line, mostly just to check the box, and had zero appetite to do anything for everything else. The rule accidentally encouraged a narrow, expensive, box-checking response.

So in the next version, he fought for a different philosophy. Something should be done on everything, just at different levels depending on how much impact it could have. He was in the room arguing the actual numbers, like how many megawatts of generation puts a plant into the tier with the heavier obligations, a line that landed at fifteen hundred. The result was a rule where your security obligations actually line up with the real risk you pose to the grid. That is the whole difference between regulation that creates security theater and regulation that creates security.

> *Low impact, low effort sets the stage for doing the right thing, because our compliance obligations line up with the risk we create.* – A power generation cybersecurity leader, on the show

And good regulation moves the entire market. That same guest pointed out that once the fines for those grid rules started landing, and the requirements became real and enforceable, it created a whole cottage industry of better security tools built specifically for the industrial world. The regulation did not just force companies to act. It pulled an entire market of solutions into existence that simply had not been there before. A guest who came up as an auditor boils the good version down to one question he asks every organization. Have you tiered your facilities? Because not every site deserves the same effort. The storage yard with nothing critical in it should not get the same spend as the plant that could hurt somebody or stop the business cold. Match the effort to the risk. That is the entire game.

Which Rules Apply to You?

Most of my scar tissue comes from the power grid, so that is where my stories live. But you might be reading this from a dentist's office or a machine shop, wondering which rules are pointed at you. So here is the quick map, no deep dives. If you hold personal information about customers or employees, privacy laws apply to you, and depending on where those people live, that can mean state rules or European ones. If you are a public company, disclosure rules now put you on a clock to tell investors about a serious incident. If you touch health information, the healthcare privacy rules follow that data wherever it goes. If you take credit cards, the payment card industry has its own security standard, enforced through your ability to keep taking cards. If you sell to the Department of Defense, there is a whole certification regime coming for you and everybody in your supply chain. And no matter what you do, every state in the country now has a breach notification law that says when you lose people's data, you have to tell them. The details change. The lesson does not. Figure out which two or three of these apply to you, treat them as your floor, and build up from there.

When the Rule Makes an Engineer the Cyber Boss

There is a human cost to all of this that nobody talks about. When a big new regulation lands, it does not come with a trained security team attached. That same researcher watched it happen when the major grid rules became mandatory. She saw a bunch of engineers, hired for their engineering skill, suddenly told, congratulations, you are now the cybersecurity person for your plant. No training, no background, just a brand-new mountain of requirements and an audit coming over the hill.

And because the rules are often ambiguous, that produces some genuinely strange work. She tells a story on herself from an early internship, right before one of these rules took effect. Nobody was sure whether a requirement to block ports meant physically or logically. So she spent part of a summer in a data center with a bag of little plastic port blockers and a spreadsheet, plugging each one in and writing down its serial number next to the serial number of the device. Hours of tedious, arguable work, done as carefully as she could, because that was the best anyone could interpret the rule to mean. That is what ambiguity in a regulation actually costs, multiplied across thousands of plants.

She did land on one genuinely useful upside of all this, though. Tying a security need to a compliance requirement is often the only way to actually get it funded. Walk into a budget meeting and say, I need a tool to protect my network, and you get a blank look and a no, because what does that even mean. Walk in and say, the regulation requires me to demonstrate this specific thing, and here is a tool that does exactly that for this amount of money, and suddenly it is a conversation that can end in a yes. Compliance, used well, is a lever for prying loose the security budget you needed anyway.

Underneath a lot of this is a translation problem. The same guest, sitting on that drafting team, realized the experts in the room were using the same words to mean different things, and could not agree because they had never agreed on the words themselves. The people on the plant floor have the same issue. Tell an operator network and adjacent network and you might get a blank stare, not because they are not smart, they are brilliant, but because nobody ever taught them the cyber dialect. Good regulation, and good security, has to be written and explained in language the people actually doing the work share.

You Cannot Buy Your Way to Either One

There is a powerful temptation, the moment a regulation lands, to go shopping. Buy the tool, check the box, declare victory. The former auditor has the best argument against that, and it comes from coaching his kid's basketball team. The head coach showed up with a whiteboard full of plays and a folder of fancy tactics, for a team of nine year olds who could barely dribble. What they actually needed was to learn to move the ball. His point about cyber is identical. If you try to solutionize your way out of it, buying tool after tool, you will spend a fortune and get nowhere. Far better to be the coach who says, let us pick two controls, execute them brilliantly, and add more next year.

He has another one I think about a lot. He is a dog guy, and he points out that no matter how many kinds of dog food the store stacks on the shelf, he still only has two dogs. He is buying one bag a month, period. The flood of new security products does not magically give a small organization more capacity to buy, deploy, and actually run them. And buying the biggest, most powerful tool before you can handle it actively slows you down. As a guy who does some track driving, he notes that too much engine in a car you cannot control gives you worse lap times, not better. You go faster in a car you can actually drive.

The deeper point is that a tool sitting in a box protects nobody. I learned this one the hard way as an asset owner. On a big project, I insisted we plan the handoff, who takes the baton and actually runs this thing after I install it, because I had been burned before. I had deployed a whole pile of technology that nobody was there to operate, so it just sat there, tuned up once and then slowly going stale. Compliance you bought but never operate is the exact same trick. It looks like security. It is shelfware with a certificate stapled to it.

Frameworks Are a Language, Not a Finish Line

So how do the good organizations actually use all this? They treat the framework, the NIST standard or the industry one, the way you treat a dictionary, not a finished book. The guest who helped write the standards makes the point that a framework is just the shared language for describing what you are trying to do. You do not pull it off the shelf and suddenly have a security program. You have to personalize it to your own business.

And here is the beautiful part. The organizations that do it right start by building the foundation for compliance, just to pass, and then they look at what they have built and realize, wait, this is actually showing me things. They start maturing past the rule, using compliance as the floor it was always meant to be, and climbing from there. The ones who do it wrong stop the very moment they pass the audit, hang the certificate on the wall, and mistake it for safety.

The asset inventory tells the whole story. I once did an assessment where the plant proudly handed me their official list of about ten assets. All ten were wrong, and I found a hundred more they did not know they had. They were perfectly compliant with a list that was pure fiction. When I told that same researcher this story, her reaction said everything you need to know.

> *I would love just to be handed an asset inventory. I would weep tears of joy, truly.* – A vulnerability management researcher, on the show

You cannot secure, or honestly even comply with, what you do not know you have. The paperwork said they were fine. The reality was a hundred unknown machines humming away in the dark.

Is Cyber Held to the Same Standard as Safety?

One question from those conversations has not let me go, and it cuts right to the heart of why compliance and security keep drifting apart. In the industrial world, safety and quality have real, hard standards. If you do not meet them, you do not get to do business. Full stop. Nobody negotiates their way around the safety standard. So ask yourself the question the standards writer asked me. Do you think cyber is held to that same standard? Or is it still the sizzle, the thing everybody nods along to in the meeting and then quietly underfunds the moment the meeting ends?

For most organizations, the honest answer is that it is not there yet. Safety earned its hard line over decades, and a lot of it was written in the wake of people getting hurt. Cyber is still climbing toward that kind of seriousness. The good news is that it is climbing. Cyber is a genuine business problem now, sitting in boardrooms it never reached ten years ago. The goal we are working toward is the day when being unable to demonstrate real security is as much of a non-starter as failing a safety inspection. We are not there yet. But we are heading that direction, and the regulations, for all their flaws, are part of what is pulling us along.

Government power in security cuts both ways, and there is one place it turns genuinely dangerous. Right now there is a real fight going on, with governments leaning on the big technology companies to build back doors into the encryption that protects all of our private messages and data, so they can read it whenever they decide they need to. I understand the impulse. But here is the problem nobody can get around. A back door built for the good guys does not stay theirs. The moment that lock is weakened, it is weakened for everybody, including every criminal and hostile government on earth. If one government can demand the keys, what exactly stops the next one? Keep giving up encryption in the name of safety, and pretty soon nobody is safe. I have nothing to hide. Most of us do not. But that was never the real question. The question is what someone could do with your private life if the lock came off, and who they could hurt through you. That door, once it is open, does not close again. It is Pandora's box, and it is a one way trip.

The Bottom Line

Regulation is not the enemy, and compliance is not a waste of time. The power grid is more secure today because of it, and a clear floor is genuinely valuable, especially for organizations that would otherwise do nothing at all. But never, ever mistake the floor for the ceiling. Passing an audit means you did the specific things on a list, for the part of your environment that happened to be in scope, in a way an auditor could check. It does not mean you are safe. The organizations that actually stay secure use compliance as a starting line, hunt down the gaps the audit cannot see, especially the ones living outside the scope, match their effort to their real risk, and keep climbing long after the certificate is framed on the wall. Do what the rule says. Then do what actually protects you. They are not the same thing, and the gap between them is exactly where your next breach is going to come from.

Tools That Help

Categories that turn a compliance program into an actual security program.

- **An accurate asset inventory.** The foundation of both compliance and security, and the one thing almost nobody actually has.
- **A governance, risk, and compliance platform.** To track your controls, your evidence, and your gaps in one honest place.
- **Vulnerability and patch management.** With the ability to track mitigations, not just patches, so the smarter fix still counts.
- **Risk tiering and business impact analysis.** So you spend the most where the impact would be greatest, and not a dollar pretending a storage yard is a power plant.
- **A security framework.** A standard like NIST, used as a shared language and a floor to build on, never as a finish line.
- **Network monitoring.** To verify that reality actually matches the paperwork, because it usually does not.
- **Compliance evidence automation.** So proving compliance steals less time from the work of actually being secure.

Your Action Steps

1. **Write down the gap.** For your own organization, what compliance requires versus what would actually make you secure. Then sit and stare at the difference.
2. **Find what is out of scope.** The vendor connections, the forgotten systems, the third parties. That is where your risk is hiding.
3. **Get an accurate asset inventory.** You cannot protect, or honestly even comply with, what you do not know you own.
4. **Tier your sites or systems by real impact**, and match your effort to the risk instead of spreading it evenly.
5. **Treat your framework as the starting line, then pick one thing the rule does not require, and do it anyway.**

Clear the floor, then keep climbing. The certificate on the wall was never the goal. Being hard to hurt was. That is the difference between passing an audit and being safe, and it is the one worth chasing.

Part IV

The World I Live In: OT and Critical Infrastructure

Chapter Eleven

Operational Technology: The Overlooked Side of Cybersecurity

The Computers That Run the Physical World

Let me tell you about a power plant, and a game of dominoes. Operators there did rounds, walking the plant on a schedule, scanning little tags mounted on the equipment with a handheld and logging the readings. Pretty standard. But the data coming back made no sense. Temperatures reading nine million degrees, numbers that could not possibly be real. So one day at lunch, I watched. An operator leaned back in his chair in the break room, reached over to a tag stuck on the wall next to the table, scanned it, set the handheld down, and went back to his game of dominoes. They had copied every tag in the field onto blank tags, mounted them all on the break room wall, timed how long the real rounds were supposed to take, and were scanning them from their chairs while they ate lunch and entered whatever numbers came to mind.

Here is what I want you to take from that, because it is the whole spirit of this chapter. Those were not bad people. They were good operators who had decided, correctly, that nobody ever looked at the round data and nothing they reported ever got fixed, so the round was busywork, and they routed around it. It was a broken people-and-process problem, not a technology problem. And that is the world of operational technology, OT for short, the computers that quietly run the physical world. It is the most important and the least understood corner of cybersecurity, and almost nobody outside of it even knows it exists.

The Computers You Never See

When you think of a computer, you think of a laptop, a phone, a server in a rack. But there is a whole other category of computer you interact with constantly and never see. The ones that run the power plant that keeps your lights on. The ones that treat and pump your drinking water. The ones that run the factory that made your car, the refinery, the pipeline, the cruise ship, the food plant. They do not have a screen for you. They do not have a keyboard for you. They just sit there, year after year, doing one job, and you only ever notice them on the rare day they stop.

And do not picture these as sleek supercomputers. A security leader I hosted, who secures cruise ships, makes a point that stuck with me. The most complex things we run are controlled by the simplest. The relays, the switches, the little controllers running a massive, intricate system are each almost dumb on their own, simple devices each doing one tiny job, wired together into something enormously complex. That simplicity is a feature, it makes them reliable, but it is also why they have no room for security built in. A device that simple was never going to ask anybody for a password. We call that world OT to set it apart from IT, the regular information technology that runs your email and your spreadsheets. And for decades, while the IT world got locked down tight, the OT world got left alone, because it was hidden, because it was specialized, and because the people who built it never imagined anybody would attack it. That is exactly why it is the overlooked side of cybersecurity, and exactly why it is now in the crosshairs.

In OT, the Rules Flip

Here is the first thing that breaks an IT person's brain when they walk into this world. Everything they know about priorities is upside down. In regular IT, the top priority is usually protecting the data, keeping it secret. If a system has to go down for an hour to install a security patch, fine, take it down. In OT, that thinking can get somebody killed.

In OT, the top priority is keeping the thing running and keeping people alive. I did not read that in a framework. I lived it for fifteen years, and it is not even a close call. One of my guests, who secures some of the largest industrial environments on earth, says it the same way I do.

> *In an OT environment, availability is the most important thing. It trumps every other priority. It is not even a close second. It is availability, a hundred percent of the time.* – An ICS and OT security leader, on the show

And the reason it matters that much is the same reason this whole field exists. These systems touch the physical world, and the physical world can hurt you.

That is not hyperbole. I have stood on outages where somebody did not go home. I will tell you about that when we get to critical infrastructure, because that is the chapter it belongs to. When you have worked in that environment, you understand why an operator looks at risk completely differently than somebody worried about an email server going down. And it is why one of the most common IT instincts, locking a screen after a few minutes of inactivity, is dangerous in a control room. I do not ever want an operator to have to log in to control their plant. I want them to be able to grab that control at a heartbeat, no password, no fumbling, no fingerprint, just reach out and stop the runaway machine. The operator who never logs in is not a security gap. In that world, it is a safety feature.

If you want to understand why operators treat risk so differently, you have to understand what they have seen. A former operator on the show talks about arc flash, where a dropped tool or a metal ring touching the wrong two points releases an explosion of energy that can vaporize a person in a fraction of a second and burn their clothes onto their skin. The protective gear is so heavy and so serious they nicknamed it after a bomb-disposal movie. When you have watched that safety video, or worse, seen it happen, an email server going down does not register on the same planet. That is the lens an operator brings to your cybersecurity proposal.

It is also why OT culture is built around safety in a way IT culture simply is not. In a control room, people use the phonetic alphabet and three-way communication, the way a pilot or a soldier does. I heard you say reboot unit one. Yes, reboot unit one. Okay, rebooting unit one. They stop the instant anybody is unsure. They hold a safety moment before a job. Our IT counterparts, bless them, are not pausing for a safety briefing before they push a patch to a Windows server on a Friday afternoon. Neither approach is wrong for its own world. They are just different worlds, and pretending they are the same is how people get hurt.

Why You Cannot Just Patch It

So the IT person says, fine, but you still have to patch these systems, install the security updates, keep them current. And the honest answer is, often, no, you cannot, at least not the way IT does. Picture a plane at thirty thousand feet with three hundred people on board. Do you want somebody patching the flight control software right now, mid-air? Of course not. You land the plane, get everybody off, and then you update it. An OT system that is running a live process is that plane, except it never lands. It runs around the clock for years.

And a lot of these systems are old. Genuinely old. It is completely normal to find a computer running an operating system that has not been supported in fifteen or twenty years, quietly running one critical step of a process. You cannot just rip it out and replace it, because replacing it can mean replacing the entire control system around it, a multi-million dollar project to upgrade one machine. So instead, the rule in this world is the one I heard on the show and adopted on the spot. You do not touch the dust. You leave the forty year old system alone, you wrap protection around it, you watch it, and you let it keep doing its one job. I compare these old systems to a car you have driven for twenty years and never once changed the oil, and the engine still has not blown. That tells you something about how these things were built. They were built to last decades, not to be replaced every three years like your laptop.

Nothing makes that concrete like the machines themselves. An OT security consultant I interviewed told me about a dairy where the single most important step of the entire process, the part that actually made the product, ran on a 386, a computer so old that most people reading this have never seen one. Every single day it overheated. The morning routine was to pull the plug, plug it back in, let it reboot, and start the process. It ran the core of that business that way for about twenty years, until they finally spent a million dollars to move it onto something modern. That is the kind of thing hiding inside the systems we are talking about protecting.

How We Got Here: Security Bolted On After the Fact

It helps to understand how this corner of the world fell so far behind. When I started focusing on this around 2010, driven by new grid regulations, there were simply no security products built for OT. None existed. So we did what you do. We took regular IT gear, the firewalls and switches and monitoring tools, and we used them differently, carefully, in a place they were never designed for. We would deploy a tool whose normal job is to block unknown devices, turn off all the blocking, and use it only to quietly tell us when something new showed up, because actually blocking something could take the plant down.

Even the smartest IT tools were lost in there. The fancy firewall that could read and classify every kind of office traffic would stare at the industrial protocols and have no idea what it was looking at, because nobody had ever taught it that language. And the sites themselves treated security as a thing you buy once. We would install a firewall and the plant would think, great, we bought a firewall, we are secure now, check the box, as if the firewall were oxygen. The good news is the industry has come a long way, and there are real OT tools now. But plenty of plants are still living in that bolt-it-on-later world, and the gap is exactly where the risk lives.

And there is one IT habit that is genuinely dangerous in OT, so much so that it has become a running fear. Scanning. In the IT world you scan your network constantly to find devices and weaknesses. One of my guests still remembers a plant security manager telling him, dead serious, that if you ran a routine network scan against one of his old controllers, it would leave a crater in the ground three miles wide. He was exaggerating, but only a little. Those fragile old devices can fall right over from traffic a modern computer would not even notice. The fear of scanning OT is not some superstition somebody invented. It is a scar. Somebody, somewhere, scanned, and something broke.

Insecure by Design

Now here is the part that should genuinely worry you. A lot of the protocols, the languages these industrial machines use to talk to each other, have no security at all. None. They were designed in an era when everything on the network was trusted, so they never check who is talking. You can send a command to open a valve, and the equipment just opens it, no questions asked. I describe it like a stranger walking into your kitchen, telling somebody to make a sandwich, and they just do it, because the guy is standing right there. No who are you, no how did you get in here.

And the inside is soft. Everybody pours their effort into the perimeter, and behind it the systems trust each other completely and talk over protocols with no locks at all. So once an attacker is past the edge, there is very little left to stop them. And it is not just the protocols. Walk into these environments and you will still find the digital equivalent of leaving every door unlocked, the kind of default passwords and wide-open services that the IT world banned fifteen years ago. One of my guests was telling a twenty-year veteran of this industry that the protocols are still wide open, and the man was genuinely offended, almost angry. You are telling me, he said, that people choose to run insecure environments? It is not that anybody chose insecurity. It is that this stuff was built in an age of trust, and the trust got baked in so deep that taking it out now means rebuilding the whole thing.

If you walked an OT network alongside an IT security person, they would have a quiet heart attack at the everyday sights. Telnet, an ancient way of connecting to devices that sends everything in plain readable text, still running. Default usernames and passwords nobody ever changed, the digital equivalent of admin and admin. Passwords sitting around in clear text. Monitoring services left wide open. A twenty year old operating system. A shared login taped to the monitor that everyone on shift uses. Every one of those was banished from the IT world more than a decade ago. In OT, they are just Tuesday.

It Is OT Because of What It Does

Here is a distinction that will save you a lot of pointless arguing. People love to fight about whether a given box is IT or OT. Is a Windows computer OT? Is a network switch OT? Is a domain controller OT? And the answer that actually matters is, it depends entirely on what the thing is doing. A network switch is just an IT product. But if that switch is the one carrying the signal that lets an operator see and control a turbine, then you had better treat it with OT-level care, because updating its firmware at the wrong moment means the operator loses visibility of a live machine. Same box, completely different rules, because of its function and its consequence.

So stop arguing about the label. I have said this for years. When something goes wrong, your CEO does not care whether the thing that broke was technically IT or technically OT. They care that the plant is down and how fast you can get it back. Look at the function the system serves and the consequence if it fails. That tells you how to treat it. The label tells you nothing.

Triage Like a Paramedic

Once you accept that you cannot do everything, and you cannot, the whole job becomes triage, the way a paramedic sorts patients at an accident. You put your effort where the impact is greatest. My favorite way to explain this is two identical computers. Same model, same software, same known vulnerabilities, the exact same risk score on paper. One of them runs the turbine that generates the power and the revenue. The other one runs the ice machine in the break room. They look identical to a vulnerability scanner. They are not remotely the same risk to the business. You do not spend your scarce time and money guarding the ice machine. You go protect the turbine.

That mindset turns an impossible mountain into a short to-do list. An ICS security instructor on the show, who teaches industrial defense, walks into water utilities staring at a hundred and fifty things they should fix, and he tells them to ignore almost all of it and do two or three things first. Build a real boundary between your IT and your OT, because more than half the time the attacker gets into OT by coming in through IT. Lock down your remote access. And get some visibility into your most critical systems. That is it. As he likes to say, this is not rocket science.

> *This stuff is not rocket science. Pick the two or three controls that matter most, and do those.* – An ICS security instructor, on the show

And the triage has to respect the operation, which means the cyber fix cannot break the plant. That same ICS security instructor and I have both seen the same well-meaning mistake. An IT or compliance person walks a plant, sees a critical computer sitting exposed, and says, we will just build a cage around it to lock it down. But in one plant, that exposed machine was the very thing operators needed to start the plant from a dead stop. You cannot cage the one thing somebody has to reach in an emergency, because if they cannot start the plant, your compliance certificate does not matter, the lights are still off. The better answer was not a cage. It was to pull the controller off the side of the machine and relocate it into a spot you could actually protect, without ever blocking the operator. The fix has to fit the job, not fight it.

OT Is the Business

I want to leave you with the idea that reframes this entire chapter. We talk about OT as if it is a niche, a backwater, the overlooked corner. But that same ICS security instructor turns that completely around.

> *If the OT part of the business did not exist, HR would not exist, IT would not exist, finance would not exist. The OT is the business.* – An ICS security instructor, on the show

Think about it. The power company exists to generate power. The water utility exists to deliver water. The factory exists to make the thing. All the email and the spreadsheets and the org charts are in service of that physical mission. The OT is not a side system. It is the reason the company has a door to walk through. Which is exactly why it is so dangerous that we overlooked it for so long, and why the people who run it, the operators, deserve to be treated as the experts they are. Go back to those operators playing dominoes. The lesson was never that they were lazy. It was that the system around them was broken, the round had no value, and the alarms screamed so constantly that everyone had long since learned to tune them out. When you fix the people and the process, with the operators, not at them, the technology finally has a chance. Get that backwards, and they will route around you every single time, because they always can.

The Bottom Line

Operational technology is the hidden layer of computers that runs the physical world, and we left it unguarded for decades because it was out of sight and out of mind. It does not play by IT's rules. Availability and safety come first, because lives and physical processes are on the line. You often cannot patch it or replace it, so you isolate and watch it instead. Its protocols were built to trust, not to verify, so the inside is soft once you are through the shell. The job is not to do everything. It is to triage by real impact, fix the handful of things that matter most, respect the operation enough that your fix does not break it, and treat the operators as the experts they are. Get that right, and you are protecting the actual business. Get it wrong, and you are guarding the ice machine while the turbine sits wide open.

Tools That Help

Categories built for the realities of OT.

- **An accurate OT asset inventory.** You cannot protect, patch, or even find what you do not know is there, and almost no plant has a real one.
- **Passive network monitoring and visibility.** Watch the traffic without touching the fragile devices, so you can see what is talking to what without risking the process.
- **Secure remote access built for OT.** Time-limited, multifactor, and brokered, so vendors and engineers can get in safely without a wide-open door.
- **Network segmentation and a real IT-to-OT boundary.** The single highest-value control, since most OT compromises come in through IT.
- **Risk-based vulnerability management.** Prioritize the turbine over the ice machine, and track mitigations, not just patches, because you often cannot patch.
- **OT-aware firewalls and monitoring.** Tools that actually understand industrial protocols, instead of staring at traffic they cannot read.

Your Action Steps

1. **Find your OT.** Walk your operation and make an honest list of every computer running a physical process. You will find more than you expected.
2. **Put availability and safety first.** Stop forcing IT habits like screen locks and surprise patches into the control room. Respect why the operator never logs in.
3. **Build the boundary.** Put a real, monitored wall between your IT network and your OT, because that is the road most attackers take.
4. **Triage by impact, not by score.** Find your turbines and your ice machines, and spend your effort where the consequence is real.
5. **Bring the operators in.** They know where the bodies are buried. Fix the process with them, and they will stop routing around you.

OT is not too complicated to secure. It is just different, and it has been ignored for too long. Pick one of these and get moving this week. This work has sat in the dark long enough, and you can be the one who finally turns the light on.

Chapter Twelve

IT and OT Convergence and the Trust Problem

We Move at the Speed of Trust

My power company bought a plant once, and the man who ran cybersecurity and compliance there absolutely despised us. He was certain that the new owners, and me in particular, were going to march in and break the systems he had spent years keeping alive. I tried everything. Donuts. Lunch. Coffee. Nothing. He stayed cold. Then one day I noticed a photo on his office wall. It was him, in a jiu-jitsu gi. I train jiu-jitsu, and I could see from the belt that he outranked me, so I said so. I told him I recognized the rank, that I was a white belt, and that I trained too. The man practically opened up on the spot. We talked jiu-jitsu for an hour. And from that day on, he saw me as a human being instead of just another suit there to wreck his plant, and he would actually listen when I wanted to change something in his environment.

Nothing about the technology changed that day. What changed was trust. And if there is one idea I want you to carry out of this chapter, it is the one that explains my entire career and most of the failures I have watched in this field. We move at the speed of trust. The faster two people, or two teams, can trust each other, the faster they can actually get anything done. And nowhere is that harder, or more important, than in the long, awkward marriage between IT and OT.

Two Tribes, One House

For most of the history of this field, IT and OT were two separate tribes who barely spoke. IT ran the email, the laptops, the corporate network. OT ran the plant. They had different tools, different priorities, different cultures, and a deep, mutual suspicion. Back when I was an asset owner, my team used to joke about printing T-shirts that said I am from corporate, or I am from IT and I am here to help, because to the people on the plant floor, those were the scariest words you could say. Whenever the corporate people showed up to help, it usually ended up hurting.

But here is the thing. Those two tribes do not actually live in two different houses. They live in the same one. An ICS and OT security leader I hosted on the show put it better than anyone I know, and it has become a rallying cry.

> *It should not be the IT side of the house and the OT side of the house. It is the same house we are all trying to protect.* – An ICS and OT security leader I hosted

I say it as the same jersey. You are IT and I am OT, and we have different jobs, the way an offense and a defense have different jobs. But we are on the same team, chasing the same goal, and if there is a guy in the corner who is the best firewall expert in the business, why on earth would the OT side not use him? The whole problem is that the jersey got torn in half a long time ago, and stitching it back together is not a technology project. It is a trust project.

Why the Mistrust Is Earned

Here is the part IT people need to hear, and it is uncomfortable. The operators' distrust is not irrational. It is earned, built up over years of help that turned out to hurt. Let me give you the example I have watched break trust more times than any other. A computer is sitting in a control room, showing an operator the live screens they use to run the plant. Somebody in IT, doing their job exactly right by IT standards, adds that computer to a policy that locks the screen after a few minutes of inactivity, or pushes out a patch. Now it is three in the morning, the operator is mid-procedure, and the screen they were watching goes dark and demands a login they have never had and do not know. They cannot see the plant. They cannot control it. And IT did not even know they had done it, or what it would cost.

That is how trust dies. Not through malice, through a reasonable IT action that was dangerous in an OT world. And it compounds, because of the other gap. IT works Monday to Friday and goes home. The plant runs around the clock. So when the thing breaks at three on a Saturday morning, it is the operator who gets the call and has to fix it, often without even knowing which IT person to wake up. After enough of those nights, the pendulum swings all the way over to, I do not want anything from you, because you cannot even get the basics right. You can force compliance up the chain if you want, but forcing an operator to accept your help is like prying a kid's mouth open to make them eat. Maybe you get one bite in. They will never learn to like it, and they will fight you forever. You might win the battle and lose the war.

You Cannot Run a Baseball Team With a Football Playbook

Part of why IT help so often backfires is that IT people arrive with the right answers to the wrong game. The problems really are the same on both sides. You have to patch things, you have to control remote access, you have to deal with old vulnerable systems. But you do not solve them the same way. I compare it to sports. You cannot take a football playbook into a baseball game and expect to win. Both need athletes, both need skill and conditioning, there are real similarities. But they are different games with different rules, and running the wrong playbook just gets you beaten. An IT team that walks into OT and runs its standard plays, patch everything tonight, scan the whole network, lock the screens, is running a football playbook on a baseball field, and the operators can see it instantly.

Underneath all of it is a truth I come back to constantly. It is people, process, and technology, and only one of those three is technology. I can buy the best woodworking tools in the world and fill my garage with them and stack the lumber right next to them, and I can come back the next morning and not one stick of furniture will have built itself. Somebody has to pick up the tools, with a plan, and do the work. Convergence fails when both sides obsess over whose tools are better and forget that the whole thing runs on people and process. The tool was never going to bridge the gap. Only people can.

Earning the Hard Hat

So how do you actually rebuild it? You earn it, the slow way, the same way you earn anything. You show up, you listen more than you talk, and you prove you understand their world before you try to change it. I have a real advantage here, because I have worn the hard hats. I have worked outages. So when I walk into a plant, I am not cosplaying, and they can tell. The sad truth is they can read the opposite just as fast. Send a kid straight out of college who has never set foot on a site, and the operators will read it off him from a mile away, from the shoes he is wearing to the way he stands. It just radiates off a person, I have no idea what I am doing here.

The reason I lean so hard on having worn the hard hats is that it lets me say the thing operators most need to hear, and actually mean it. I am not the enemy. I am more on your side than the IT folks are, because I have sat in your chair, and I am going to fight to make sure nobody breaks your stuff. When an operator truly believes that, the whole relationship changes. You stop being the corporate threat and become the person in the room who has their back.

Credibility opens doors that donuts cannot. An OT security consultant I hosted, who used to be a controls engineer, was doing a security audit, and the engineers were giving him the standard do-not-touch-my-stuff treatment. Then he leaned over at a problem they were stuck on and said, oh, that is your controller doing that, and by the way your code is a mess, you guys should have a standard. They blinked and said, I thought you were the cybersecurity guy. He told them he had programmed that exact brand of controller thousands of times back in his controls days. And just like that, the whole room changed. Suddenly he was all right, somebody who had seen things, and they wanted to know what he thought. That is the moment you are after, and you cannot fake your way to it.

> *Right away they decided I was all right, that I had seen real things, and then they opened up and asked what I wanted to know.* – An OT security consultant I interviewed

When I could not be in the room myself, I hired the credibility. I once brought on a thirty year veteran, a former operator and instrumentation technician who knew every plant, every system, every person, and had programmed every controller. He had instant standing in any control room in the company, and he could tell my team in advance, here is the resistance you are going to get at this plant, and here is how we talk to them so they will actually hear us. That kind of insider trust is, honestly, priceless.

And when somebody on your team breaks the trust, you fix it immediately and in person. An IT hire of mine once rebooted a machine at a plant remotely, from four hours away, on a Friday afternoon. It came back up fine, and he figured no harm done. I told him to get in his car, drive the four hours, walk up to that operator, explain exactly what he did and why, and sit there until everyone was satisfied things were okay. He protested that it had come back up. I told him that did not matter. If we do not do this, we have burned a bridge, and they will never let us touch their stuff again. Sometimes you have to eat crow and drive the four hours.

Meet Them Where They Are

Rebuilding trust is not all grand gestures. A lot of it is small, practical, and human. The OT security consultant who used to be a controls engineer has a perfect example. Operators are not supposed to bring phones into the control room, but the game is on Sunday and they are going to smuggle one in anyway. So rather than fight a losing battle, he makes sure there is a powered charging plug right there on the control room desk. Why? Because if there is nowhere to charge the phone, the operator will plug it into the control system to get power, and now you have a real problem. Meet the human need, and you remove the risk.

The same instinct shows up in how the best people build a room. One of my guests walks into a site with a stack of donuts, gets everybody, IT and OT, around a whiteboard, and starts mapping out how everything actually connects. And there is always a magic moment when he asks, so how does this part talk to that part, and the whole room goes quiet and looks at each other, because nobody actually knew. That shared, slightly embarrassing realization is the point. Now they are solving it together, on the same side of the table. An ICS security instructor I had on the show describes the same idea as brisket, bourbon, and barbecue. You build the relationships over a shared meal, with no agenda, and those relationships are exactly what save you when the crisis hits. Or as my dad always said, two ears and one mouth. Listen at least twice as much as you talk.

And getting everyone around that whiteboard does something else, it quietly drags the hidden risks into the light. Time and again the same scene plays out. A plant insists it has no remote access at all, and then somebody traces a line on the board and says, well, except for that, and points to a cellular card plugged into a machine in the back that runs straight out to a vendor, with shared passwords and no logging, a modem that sits open all the time. Nobody was lying. They genuinely did not think of it as remote access. Convergence done well is partly just this, getting the two halves of the house in one room until the whole map of reality finally appears on the wall.

Active Directory and the Dry Brush

Now let me show you what the trust problem looks like when it goes wrong technically, because there is one place IT and OT convergence quietly becomes a catastrophe. Active Directory, the system that manages logins across a network. In the IT world, you connect everything to one central directory so it is easy to manage. The instinct is to extend that same directory right into the plant. And an ICS and OT security leader I hosted says he hears the same confession almost weekly. A company shared one directory across

IT and OT, ransomware got into the IT side, and it spread straight into the plant like it was built to, because it was. It took everything down. He figures something like one in five environments still does this.

> *It is acres of dry brush waiting on a single spark, one person clicking one link, and then it races through IT and OT alike.* – *An ICS and OT security leader I hosted*

It gets worse. When a vendor sets up that directory in the plant, the account they leave behind is very often the most powerful kind of account there is, able to control everything, and everybody logs into every machine with it. An attacker just sits and waits for somebody to use those god-mode credentials, and then they own the place. I have seen a shared directory literally trip a running power plant, a vendor's mistake cascading into a unit coming offline. This is exactly why I never, ever commingle IT and OT directories.

But, and this is the trust lesson, the answer is not for OT to lock IT out entirely. The answer is the general contractor model. A good general contractor is knowledgeable enough to look at the electrician's work and say, that is not up to standard, but he does not do the wiring himself, because he is not the craftsman. So lean on your IT team that manages directories for a living. Have them come look at your environment and advise you, because just because the vendor set it up does not make the vendor an expert. IT brings the deep skill. OT keeps its environment separate and keeps the final say. That is convergence done right, two experts, one house, clear boundaries.

Building the People Who Bridge the Gap

There is a reason this gap is so hard to close. The people who genuinely understand both worlds are rare, and you do not find them, you grow them. I learned that running a tiny OT team, six of us covering more than forty power plants, including a nuclear plant and mines, while the IT side had hundreds of people and ten times the budget. I could not hire my way out of that. So I grew my own. I brought on people with no OT background at all, and after their onboarding training, they never once went to their office. They went straight to outages. For the next six months they lived at power plants, sixteen-hour days, seven days a week, learning the real thing by doing the real thing. Six months later they were experts, rock stars, in environments they had never even seen before they started.

One guest, who came up in the Navy, has the perfect framing for it. Smooth seas never made skillful sailors. You build the people who can bridge IT and OT by throwing them into the hard thing and letting them learn to steer. You do not get a sailor by keeping the ship tied up in the harbor. And this matters more every single year, because the veterans who built these plants, the ones who know where every wire runs, are retiring, walking out the door with decades of knowledge in their heads. The most valuable thing you can do for convergence is take a curious IT person and a willing operator and give them the chance to learn each other's world before that knowledge is gone for good.

When the Trust Pays Off

Let me tell you what earned trust actually buys you, because it is the most rewarding part of this whole job. By the time my team had supported those plants for a few years, the relationship had completely flipped. Operators and plant managers started calling us, not to complain, but for help. A vendor would pitch them some new equipment, and the plant would phone my team and say, these folks want to sell us these firewalls, but I know you guys do firewalls, can you sit at the table and ask them the right questions so we pick the right thing? Read that again. The same people who once wanted to print T-shirts to keep us away were now asking us to sit beside them and protect them. That is what trust earns. You stop being the threat and become the trusted advisor.

And the good news is the whole industry is moving this direction. You see it at the conferences. Events that used to be strictly OT now have IT people in the same room, something that would never have happened a decade ago. One of my guests even got OT written into a government technology panel that had not included a word of it, and the whole call agreed without hesitation, something that simply would not have happened ten years back. The walls are coming down. If you want to speed it up in your own organization, here is my favorite prescription. Send your IT team to work an outage. Put them on the floor at two in the morning when something breaks, and let them feel why the operators are so protective. They will come back changed, and half the battle will already be won.

One last tactic, because it captures everything. The fastest way I ever earned trust was not by talking about cyber at all. The first real thing our monitoring ever did at one plant was hand the operators a reliability win. It caught a hidden failure they had no idea was there, the kind of quiet problem that bites you in the middle of the night, and it surfaced in minutes. We did not sell them cybersecurity that day. We handed them reliability, the thing they actually cared about. And after that, they trusted us with the cyber, because we had proven we were on their side.

The Bottom Line

IT and OT convergence is not really a technology problem. It is a trust problem wearing a technology costume. The two sides spent decades as separate tribes, and the mistrust the operators carry is earned, built from years of well-meaning IT help that broke their world. You do not fix that with a tool or a mandate. You fix it the slow, human way. Show up, listen more than you talk, prove you understand their world, fix your mistakes in person, and meet people where they actually are. Get the hard technical boundaries right, especially keeping your directories and your networks separate, so a spark on the IT side cannot burn down the plant. Do that, and IT and OT stop throwing things at each other and start playing for the same team, in the same house, in the same jersey. Because in the end, we move at the speed of trust, and trust is the only thing that makes any of the rest of it possible.

Tools That Help

Categories that support a healthy IT and OT marriage.

- **Separate identity and access management for IT and OT.** Do not share one directory across both. It is the dry brush waiting for a spark.
- **Secure, brokered remote access.** So vendors and IT can help without a permanent open door into the plant.
- **Network segmentation and a monitored boundary.** The wall that lets the two worlds exchange what they need without one infecting the other.
- **Joint tabletop exercises.** Get IT, OT, and safety in one room to practice together, which builds the relationships before the crisis.
- **A shared asset inventory and OT-aware monitoring.** A common picture of reality that both teams can actually trust and talk about.

Your Action Steps

1. **Send your IT team to work an outage.** Let them feel why the operators are so protective. Nothing builds empathy faster.
2. **Separate your directories and networks.** Make sure a ransomware spark on the IT side cannot jump straight into the plant.
3. **Lead with goals, not prescriptions.** Walk in asking how can we solve this together, not here is what you are doing wrong.
4. **Earn it in person.** When somebody breaks trust with a plant, fix it face to face, even if you have to drive four hours.
5. **Build the relationship before you need it.** Break bread with no agenda, because the trust you build over barbecue is what saves you in the incident.

Convergence is coming whether the two tribes like it or not. You get to decide whether it happens with trust or with a fight. Pick one of these and begin this week.

Chapter Thirteen

Critical Infrastructure and the Case for Resilience

Lives, Not Data

Let me start with a cookie. In a food plant, there is a peanut area and a non-peanut area, kept carefully apart, because for a person with a peanut allergy, cross-contamination is not an inconvenience, it is a coffin. One day a disgruntled employee crossed from the peanut side to the non-peanut side. The contamination was not caught until much later, downstream, when quality testing found peanuts in a cookie headed to a major retailer, a cookie that should never have contained a trace of them. The company lost that retailer. The brand took a catastrophic hit. And when a security leader I interviewed, who works in food and agriculture security, described that scenario to a food defense professional, his answer came back flat and immediate. On his report, that gets classified as terrorism. Specifically, domestic terrorism.

I open with that because the badge readers, the cameras, and the access controls that should have stopped that employee, and did not, because some were missing, are a cybersecurity problem. And the thing they were protecting was not data. It was human lives. That is the entire point of this chapter. Critical infrastructure, the power and water and food and ships that society runs on, is where cybersecurity stops being about information and starts being about whether people get hurt. It raises the stakes higher than anything else in this book.

A Bad Day Here Means Somebody Does Not Go Home

In the regular world of IT, a bad day means lost data, lost money, a rough week of cleanup. In critical infrastructure, a bad day can mean somebody does not go home. A security leader I interviewed puts the difference in the bluntest possible terms. IT is about data. OT is about safeguarding lives, because we could kill somebody. A maritime cybersecurity leader I spoke with, who secures cruise ships, says the same thing from the deck of a ship. If a vessel loses propulsion or control, it does not just inconvenience people. It can put everyone, on the ship and off it, in real danger. He points to the cargo ship that lost electrical power, lost control, and struck a bridge in Baltimore. That was a mechanical failure, not an attack, and that is exactly his point. Whether the ship's systems fail on their own or because somebody made them fail, the physics are the same, and preventing that outcome is what this field exists to do.

An ICS security instructor I hosted, who teaches industrial defense and works across water, power, and oil and gas, captures the weight of it in a single line that has stuck with me. He is describing what defenders are up against, attacks that can kill people and destroy equipment you cannot simply order a replacement for, and then he lands it with a dark little shrug.

> *It means injury, and possibly loss of life. But it is only critical infrastructure, after all.* – An ICS security instructor on the show

That sarcasm is the whole problem in one sentence. The systems that matter most are the ones we have protected least. And the numbers back up the worry. The ICS security instructor points out that, by his count, something like a fifth of industrial facilities has been compromised, and a compromise here does not mean somebody read your email. It means the control system itself was impacted, the thing that actually runs the physical process.

The food and agriculture security leader carries a question that a food safety expert once asked her on air, and it haunts me. How many bodies need to be on the floor before somebody does something? Her honest answer was, probably a lot. And then she learned that the person who asked it had personally lost a child to a foodborne illness. That is the real weight of this work. Not a compliance finding, not a quarterly metric. A parent who got a phone call. When you keep that in view, the urgency of all of this stops being abstract.

I do not know this from a podcast. I have been at a plant during an outage, and I have been around safety incidents, including one where a person did not go home. Everything about how these facilities run flows from that reality. Every morning starts with a meeting to understand everything happening on site that day. Every job begins with a pre-job brief that pulls in every person and every system owner the work could touch. We lock out and tag out equipment so nobody can energize a line while someone is holding it, because the breaker you flip at home while installing a ceiling fan gives you a jolt, and the same mistake in a substation gives a family a funeral. There is stored energy everywhere, steam under pressure, heavy equipment, long falls. The hard hat exists so a tape measure dropped from the eighth story gives you a chance. The steel toes exist so your foot is not crushed. Every piece of that gear is a lesson somebody paid for. When you have stood in that world on its worst day, lives, not data, stops being a slogan. It is just the truth.

And if you want the clearest proof that lives are the target, it has already been tried. A few years back, attackers got into a petrochemical plant overseas and went after the one system whose only job is to keep people alive, the safety system that shuts everything down before a failure can hurt somebody. The only reason nobody died is that the attackers made a mistake and tripped the plant by accident. They were not after data. They were after the last layer between a running process and a funeral.

The Sectors You Never Think About

Critical infrastructure is not one thing. It is a dozen hidden worlds, each one more fragile and more vital than the public realizes. Take a quick tour with me.

The power grid. The most interconnected machine ever built, and one with a strange secret. To start a power plant from a dead stop, you need power, which is the chicken-and-egg problem of the black start. Utilities keep careful playbooks for it, starting a small plant to make just enough power to start the next one, and phasing them in across the grid. If any link in that chain fails, you are stuck in the dark. We have never had a true nationwide blackout, and the people who keep it that way lose sleep over the day we might.

Water. The sector that worries me most, and the one that never gets any love. Small water utilities often have no real boundary between their networks, no incident response plan, sometimes barely any staff, and unlike the power grid, no regulation forcing them to fix it. They are running the most essential service there is, clean drinking water, on a shoestring. As the same instructor says, the cyber piece is not separate from the mission. A utility can spend its whole budget on new pipes, but it gets nothing if the water comes out at the wrong pressure or poisoned.

Food and agriculture. A world most of us never think about as infrastructure at all. It only joined the official list of critical infrastructure sectors in the early 2000s, an afterthought compared to power and water, and the food and agriculture security leader finds how little attention it gets, even now, genuinely disturbing. When a major meatpacker got hit with ransomware, she walked through the ripple effects nobody plans for. If the plant cannot process, where do grocers buy meat? And the cattle that could not be slaughtered are now standing in trailers and holding pens, with nowhere to go, because the farm has no feed or space left for them either. Nobody tested a backup plan, because everybody assumed business as usual. And food, she warns, is a weapon. A hostile nation that wants to break a country goes after its electricity and its food, because they want you to freeze in the winter and starve in the dark.

A friend of the food and agriculture security leader, an agricultural futurist, once tried to go ten days without food as an experiment and could not finish it. It led him to a darker question. How many hours would it take you to commit a crime if your family was hungry and there was simply no food to buy? His estimate of the average was about eight. Eight hours from empty shelves to chaos in the street. We are, as she puts it, a breath away from a foodborne crisis, because we do not understand the technology now woven through the systems that feed us. And the dependencies only deepen. Water utilities, desperate for revenue, are now selling their treated water to the data centers springing up next door, which means the industrial systems of one critical sector are getting wired directly into another. Your cloud, it turns out, needs OT too.

Maritime. The maritime cybersecurity leader corrected my whole mental model of a ship. You picture one big factory. It is actually a constellation of thirty to seventy smaller factories, all running at once. It makes its own drinking water through reverse osmosis. It treats its own sewage, so cleanly that the water it discharges is cleaner than what it took in. It generates and manages its own power, runs its own navigation, its own life support. A single cruise ship is a floating city, and a major line might run dozens of them, which he reframes as dozens of cities at sea, each one with its own customs, its own laws at every port, all of it depending on those simple little controllers never failing.

> *A ship can get by without cyber, but I cannot get by without the ship.* – *A maritime cybersecurity leader I interviewed*

And here is a detail about the grid that most people never learn. The reason Texas runs its own electrical grid, largely separate from the rest of the country, is not stubbornness. The separation began decades ago so Texas utilities could stay out of federal regulation. But the structure has a side effect engineers now lean on. A cascading blackout in one region cannot roll across the whole nation at once, and the regions can help restart one another.

Bringing that grid back up after a total collapse is sequenced as carefully as the power-up procedure on a crippled spacecraft, where doing things in the wrong order trips everything again because there is not enough power to go around. That is the invisible, fragile choreography keeping your lights on, and every piece of it now runs on ordinary computers. As the maritime cybersecurity leader points out, IT and OT alike run on the same machines you could buy at any electronics store, with all the same vulnerabilities, just doing a far more dangerous job.

How Fragile It Really Is

The thing that ties all these sectors together is how breathtakingly fragile they are underneath, and how blind we are to it. We connected this infrastructure to networks before we ever secured it, the way you might hand a trusting grandparent a smartphone with no guardrails and then act surprised when they fall for the first scam. The guardrails were never put up at the beginning, and now we are trying to add them to systems that are decades old and still running.

Two plants that start out identical do not stay that way either. An OT security consultant I interviewed told me about a power company and a financial company that each bought the exact same plant, same turbines, same drawings, from the same engineering firm. Over the next twenty years, the power company kept upgrading its control systems. The financial company, for whom the plant was just a side investment, did nothing at all. Two decades later, one of them was modern and the other was a museum of twenty year old technology that somehow still ran. They called it the rich twin and the poor twin. When you wonder why one company recovers from an attack in an hour and another takes a month, this is a big part of the answer. They did not start in the same place, and they sure did not maintain it the same way.

And people will always route around a control that gets in the way of the physical job, just like the operators playing dominoes a couple of chapters ago. The food and agriculture security leader tells of catching a maintenance manager on a scissor lift, physically moving a wireless access point, because forklift drivers could not get a signal to pick orders. The reason? She had learned on the job that a wireless signal does not pass through stacked bags of flour, and a full warehouse was blocking it. Nobody had mapped that. They just worked around it. That is critical infrastructure in real life, brilliant people doing whatever it takes to keep the product moving, security be damned, because the product moving is the whole job.

An OT security executive on the show, who sees the inside of thousands of these environments, does not sugarcoat the state of them.

> *These sites are not resilient, and they are not resilient because the network underneath them is a mess, and we have let it stay that way.* – An OT security executive on the show

The maddening part is that the gaps are usually discoverable, if anybody actually looks. The ICS security instructor runs realistic tabletop exercises at water and wastewater utilities, sitting the engineers on one side and the IT leaders on the other and walking them through a believable attack. Time after time, a utility that swore up and down it had an air gap, a complete separation between its networks, discovers in the middle of the exercise that it does not, that there is a path running right through. He insists on having the engineers in the room, because that is the moment the learning happens, in real time, with high-fives going around as people finally see their own environment clearly. The gaps were always there. Nobody had ever sat down together and traced them.

The Hidden Attack Surface

If you could see what is actually sitting on these critical networks, it would keep you up at night, and sometimes it is almost funny. The OT security executive's company watches tens of millions of devices, and he runs a little guessing game about what is hiding on industrial and hospital networks. Game consoles, tucked into break rooms and on the same flat network as life-critical systems. Smartwatches by the tens of thousands, a huge share of them so old they stopped getting security updates years ago, walking in and out on people's wrists every day. Personal vehicles, pickup trucks and electric cars, quietly connected to the plant Wi-Fi. None of it on a guarded guest network. All of it a door.

It got worse during the rush. When everyone built new distribution centers during the pandemic, there was not enough automation equipment in the world to go around, so every site became a science experiment, two centers built the same month by the same crew coming out completely different. Now the race is to build AI data centers as fast as physically possible, chasing what the industry calls time to power, stripping whatever gear they can get their hands on and bolting it in. When you build that fast, the boring step always gets skipped, the one where you verify the thing, change the default password, close the back door. So the back doors stay open, by the thousands, in the infrastructure we depend on most.

The Silver Tsunami

There is one more threat to critical infrastructure that no firewall can stop, and it is demographic. The people who built these plants, who know where every wire runs and why that one valve sticks and how to start the unit when the computer is lying to you, are retiring. All at once. The OT security executive wrote a paper about it with a title I love, calling that veteran operator the most dangerous man in the plant, because when he walks out the door for the last time, decades of knowledge that lives nowhere but in his head walks out with him.

We are losing the human resilience that has quietly held this stuff together for years, the old-timer who could walk up to a balky pipeline with a hammer and a cigarette and know exactly where to tap it because he had been there since 1958. You cannot download that. You can only transfer it, person to person, before it is gone. Which means closing the workforce gap is not a nice-to-have for critical infrastructure. It is a race against a clock that is already running, and the most valuable thing a plant can do is pair its retiring veterans with curious newcomers while there is still time.

Resilience Is the Whole Game

So how do you protect something this vital, this fragile, this complex, and this underfunded? You already know from chapter eight that the real answer is resilience, not a perfect wall. Here the stakes go up. In chapter eight, resilience meant your business survives the bad day. In critical infrastructure, it means the lights stay on and people go home. This is the single most important shift in the chapter, and it is the hill I will die on.

> *We keep cycling through recovery mode. We have to break out of it and start building for resilience, because the next event is coming.* – A security leader I interviewed

Resilience means assuming the bad day will come and engineering so you can take the hit and keep going. Maritime is resilience in physical form. The maritime cybersecurity leader points out that no matter how much you digitize a ship, everything is still physics, and you can always go down into the machinery space and press the button by hand. There is always a manual fallback, always redundancy, always a spare part, because at sea you cannot call anybody for help. That is the mindset critical infrastructure needs everywhere. When the 2024 software outage hit, a friend of the food and agriculture security leader at a cheese company in Wisconsin barely flinched, because they had backup systems, rolled the bad update back overnight, and never lost production. They had focused on resilience more than recovery, and that is the whole idea in five words. The OT security executive's whole team now flies that flag, half-joking that good security is just IT with good fundamentals, and the real banner is resiliency.

But resilience is not something you can buy either. I would rather you run mediocre technology really well than the best technology badly. I tell people the fanciest platform in the world is a Ferrari, and if you take the wheels off and pour diesel in the tank, it is going nowhere. Resilience lives in people and process far more than in any product. And part of it is the discipline to ask, before you connect one more thing, do we actually need this? Every sensor you wire up so you can watch a tank temperature from your couch is also one more door into a plant that runs around the clock anyway. Sometimes the most resilient choice is the piece of technology you decide not to add.

And the threats are not theoretical. A coordinated attack on a country's power grid did real, physical damage, bricking the equipment that keeps the lights on. I am going to walk you through that whole attack in chapter fifteen, because it deserves more than a paragraph. The adversaries are rehearsing on the infrastructure we all depend on. The ICS security instructor's advice is the only sane response to that.

> *I would not wait around for a mandate, because the adversaries are not taking any time off.* *– An ICS security instructor on the show*

Where to Start

If you run any piece of this, the mountain looks impossible, so do not try to climb all of it. A friend of the food and agriculture security leader in the UK gave her the perfect framing. Stop trying to boil the ocean. Sit down, make a cup of tea, and focus on the smaller, cheaper, people-and-process things first. The ICS security instructor's triage from chapter eleven, the IT-to-OT boundary, remote access, and visibility, is exactly where that first cup of tea should be spent.

And learn to ignore the noise so you can hear the signal. In every control room I have ever worked, the one key that physically wears out and breaks on the keyboard is the mute button, because at three in the morning the operators are hammering it over and over as alarms scream about things everyone already knows about and nobody has fixed. When everything alarms, the alarms mean nothing, and no human and no system can act on a flood. Resilience starts with deciding what actually matters, the handful of systems that can truly hurt you, and pouring your limited attention there. You will never protect all of it equally. The whole art is choosing well.

And if you are one of those shoestring utilities, hear this. You are not alone, and some of the best help is free. The national cyber defense agency, CISA, runs free vulnerability scanning and assessments for critical infrastructure. The water sector has its own information sharing center, and your state and sector associations run programs built for utilities exactly your size. The help exists. You just have to pick up the phone and ask for it.

The Bottom Line

Critical infrastructure is where cybersecurity stops being about data and starts being about whether the lights stay on, the water stays clean, the food stays safe, and people make it home. It is the most vital part of the digital world and, for decades, the least protected, run on aging systems, thin budgets, and the irreplaceable knowledge of people who are now retiring. You cannot make it perfect, and you will drive yourself crazy trying. What you can do is build for resilience, assume the hit will land, keep a manual fallback and a tested recovery, close the boundary between IT and OT, and pour your scarce attention onto the handful of systems where the consequence is measured in lives. The adversaries are already rehearsing. The only real question is whether we get serious about this before the bad day, or after it.

Tools That Help

Categories that build resilience into critical infrastructure.

- **An accurate asset inventory.** You cannot protect, or recover, what you do not know is on your network, including the game console and the smartwatch.
- **Network segmentation and an IT-to-OT boundary.** The highest-value control, since most compromises cross over from IT.

- **Secure remote access.** Time-limited, multifactor, and brokered, so vendors and staff can help without leaving a door propped open.
- **Network visibility and monitoring of critical assets.** Watch the traffic to and from the systems that actually matter, without touching the fragile devices.
- **Tested backups and recovery.** Proven restores and the ability to roll a bad change back overnight, the way the resilient ones do.
- **Documented manual fallback procedures.** The pen-and-paper, press-the-button-by-hand plan for when the technology is gone, kept current and practiced.
- **Tabletop exercises with engineering in the room.** Because the gaps only show up when IT, OT, and operations work a real scenario together.

Your Action Steps

1. **Treat it as lives, not data.** Let the stakes set your priorities. Protect the systems where a failure can hurt someone, first.
2. **Pour your tea.** Pick two or three things, the IT-to-OT boundary, remote access, and visibility, and actually do them, instead of drowning in a list of a hundred and fifty.
3. **Build a manual fallback.** Make sure you can run, even badly, when the technology is down, and practice it before you need it.
4. **Capture the knowledge before it walks out the door.** Pair your retiring veterans with newcomers now. That window is closing.
5. **Fix what the alarms are screaming about.** Quiet the noise so the real signal can be heard, and put your attention on the systems that can actually cause harm.

This is the part of the field that matters most, and the part we have neglected longest. The good news is that the answers are clear and within reach. Pick one and start this week. You do not need permission and you do not need perfection. You only need to begin.

Part V
The Road Ahead

Chapter Fourteen

AI and Cybersecurity: The Double-Edged Sword

Why You Cannot Afford to Sit This One Out

I want to start this chapter by admitting something. For a while there, I got a little tired of the letters A and I. You could not walk the floor of a security conference without every single booth slapping AI on a banner. The same firewall they sold you last year, now it is AI powered. The same dashboard, now with AI baked in. Half of it was real and half of it was a sticker. And it bugged me, because the hype got so loud that it drowned out the actual conversation we needed to be having.

But here is the other side, and it is the more dangerous side. While a bunch of folks were rolling their eyes at the hype, a bunch of other folks were quietly deciding to just avoid the whole thing. Too risky, too new, let somebody else figure it out. And that, to me, is the bigger mistake. Because this is not a sticker. It is real, it is here, and it is already changing how both the attackers and the defenders do their jobs. You do not have to love it. But you cannot afford to hide from it.

Not a True Believer, and Not a Skeptic Either

There are two easy ways to be wrong about AI. You can be a true believer, the person who thinks it is magic and wants to point it at everything. Or you can be the obsessive skeptic, the person who has decided it is all nonsense and refuses to touch it. Both of those people are going to get run over. A former government cyber operator I hosted on the show, the same operator I quoted back in chapter two, put the right posture better than I could.

> *Our job is not to be true believers, but it is also not to be obsessive skeptics.* – *A former government cyber operator*

That is the whole spirit of this chapter. I am not here to sell you on AI, and I am not here to scare you off it. I am here to help you use it without getting burned, and to make sure you are not the one left standing on the sidelines while everybody else figures out how to do more with less.

Why You Cannot Afford to Avoid It

Let me make the case for using it, because it is a strong one. AI is a force multiplier. The teams that are winning right now are not the ones with the biggest budgets. They are the ones using these tools to do more with the people they already have. And if you are a small shop, or an under-resourced team, or a department of one, that leverage is the difference maker. It does not replace a skilled person. But a skilled person who knows how to use AI will run circles around one who refuses to.

> *There are two kinds of people now: the ones learning to work with AI, and the ones who are out of a job.* – *A cybersecurity media host*

That one is a little blunt, and maybe a little harsh, but the direction is right. I am not worried about AI taking your job. I am worried about the person who took the time to learn AI taking your job. The good news is that learning it is not hard. You do not need a computer science degree. You need curiosity and a willingness to pick it up and try.

Give Your People a Superpower

Take data. Every business is drowning in it and starving for answers. The old way, you put in a ticket, you wait for the one analyst who knows the database, and three weeks later you get a spreadsheet that is already out of date. The new way, somebody on your team describes what they want in plain English, points the AI at the data, and has the answer in minutes. And not just a number. They can ask it to find the patterns they did not even know to look for. The customer who always pays late. The sensor that drifts right before a failure. The week where everything quietly went sideways.

Used well, AI takes the person you already have, the one who knows your business cold but is not a programmer, and gives them the reach of a whole team.

What It Is Actually Good At Right Now

AI is too big a word. It means a hundred different things, and that is part of why the conversation gets muddy. So let me get concrete about where it genuinely earns its keep today, in security and in everyday work.

- **Turning a wall of technical noise into plain language.** Logs, alerts, a flood of data, and it hands you a readable summary. That alone is a gift for the first pass of triage in a security operations center, where analysts drown in alerts.

- **Drafting and summarizing.** Policies, reports, emails, meeting notes. You still read it and fix it, but it gets you off the blank page, which is half the battle.
- **Writing small bits of automation and scripts.** Even if you are not a coder, it lowers the bar to automating the repetitive stuff that eats your day. Free automation tools paired with AI let a small team punch way above its weight.
- **Research and learning.** Ask it to explain a concept five different ways until it finally clicks. It is a patient tutor that never gets tired of your questions.
- **Running tabletop exercises.** Some of my guests use AI to generate realistic incident scenarios on demand, so a team can practice its response far more often than the old once-a-year drill. And practice is the whole point.

> *Anything you only practice once a year is not good enough.* – *An OT security author and trainer*

Notice that none of that is science fiction. It is the boring, repetitive, time-eating work that burns your people out. Hand that off to the machine, and free your humans up for the judgment calls, which is where they are actually worth something.

The Other Edge of the Sword

Now the part that keeps you honest. The same tool that helps you is helping the bad guys, and it is also a brand-new way to get attacked. Here is how AI cuts you.

- **It makes the attacks better.** We already talked about this. Flawless phishing, deepfakes, cloned voices, all at scale. The attackers adopted AI faster than most defenders did.
- **It lies with total confidence.** AI will hand you a wrong answer in a calm, authoritative voice and never blink. People call these hallucinations. Never trust it blindly on anything that matters. Treat it like a smart intern who sometimes makes things up, and check its work.

- **It leaks what you feed it.** Whatever you type into a public tool may not stay private. Do not paste your secrets, your customer data, or your crown jewels into one. The rule from earlier in this book still holds. Do not put anything into it that you would not be comfortable posting on social media.
- **Your own AI can become the way in.** This is the one most people never see coming. An attacker hides instructions inside a document or a web page, your AI reads it while trying to be helpful, and it quietly does what the attacker said. We have already seen real cases where a company's own AI assistant could be turned into the leak.
- **Shadow AI.** Your people are already using these tools, with or without permission, pasting company information into whatever free site they found. You cannot govern what you refuse to acknowledge.

> *Automate a bad process and you just make it bad faster, and the same goes for an AI agent.* – A security executive

What You Feed It Is the Whole Ballgame

Here is the rule that ties all the risk together. With AI, where you run it and what you feed it matter more than almost anything else. The exact same tool can be perfectly safe or a quiet disaster, and the difference is the data you put in and the place you put it.

Start with personal information. PII, personally identifiable information, is the stuff that points to a real human being. Names, Social Security numbers, addresses, birth dates, account numbers, health records. The second you paste that into a public AI tool, you have lost control of it. You do not know if it gets stored, if it trains the next version of the model, or if it turns up in somebody else's answer six months from now. And if that is your customers' data, you did not just make a mistake, you may have broken a law. The regulators do not care that it was a chatbot.

Now your intellectual property, your secret sauce. Your source code, your designs, your pricing, your contracts, the thing that makes your company yours instead of your competitor's. I have watched well-meaning people paste a confidential document into a free AI tool just to get a quick summary, and in that one click they handed the crown jewels to a system they do not own and cannot get them back from. You would not email that file to a competitor. Dropping it into the wrong AI is not all that different.

So before anybody on your team uses one of these tools, they need to answer two questions. What is this data, and where am I about to send it. Public data into a public tool, great, go fast. Sensitive data only goes into a tool that runs privately, inside your own walls, where nothing leaves. That one habit, classify before you feed, prevents most of the AI disasters I see.

You Cannot Automate Clarity You Do Not Have

This is the line I keep coming back to, because it is the one people most want to skip. AI is not magic. It runs on your data and your processes, and it depends on clean data, stable architectures, and known baselines. If those are a mess, AI does not fix the mess. It just makes a faster, more confident mess.

So before you bolt AI onto your business, do the boring fundamentals first. Know what you have. Clean up your data. Understand what normal looks like, so the AI has something true to learn from. AI on top of a solid foundation is a rocket. AI on top of chaos is just faster chaos. Do the unglamorous work first, then add the power tool.

Treat AI Like an Employee, Not an Oracle

Here is the mental model that keeps you out of trouble. Do not treat AI like an all-knowing oracle that you obey. Treat it like a new employee. A fast, eager, talented new hire who is also a little naive and will absolutely do something dumb if you hand them too much access on day one.

> *We should treat the AI a bit like a person. Would you trust someone who has access to everything? – A cybersecurity advisor*

You would never give a brand-new hire the master password to everything on their first morning. So do not give it to an AI either. You onboard it the same way you onboard a person, with the least access it needs to do the job and not one bit more.

- **Give it the narrowest access that works,** and nothing extra. Point it at the one folder or the one system it needs, not your whole drive.
- **Keep a human in the loop for anything you cannot undo.** Sending money, deleting data, making a change in production. A person approves the irreversible stuff.
- **Do not give it standing credentials.** One of my guests, who builds guardrails for AI agents, lives by the idea of starving them of

permanent access. Use keys that expire and rotate, so a tool that gets compromised cannot do lasting harm.

- **Log what it does.** So you can see what happened, answer for it later, and catch it when it goes sideways.
- **Write down the rules.** Even a one-page policy of what is okay and what is off limits puts you ahead of most organizations.

When you boil all of this down, the thing that actually keeps you in control is identity and access management. That is the unglamorous discipline of deciding who, and now what, is allowed to touch which systems and which data, and being able to prove it. We have done it for people for years, badges and logins and permissions. AI is forcing us to finally do it for machines too.

Every AI tool and every agent is now an identity living in your environment, and it has to be treated like one. It gets its own account, not a borrowed human one. It gets the narrowest set of permissions that still lets it do the job. Its access gets logged and reviewed. And the moment it no longer needs something, that access goes away. If you cannot answer what your AI can reach and what it has touched, you are not in control of it. You are just hoping. And hope is not a strategy.

Barricades, Not Promises

Here is the piece most people miss. It is not enough to control AI at the identity layer, the question of who it is and what it is allowed to log into. You have to control it at the action layer, the actual thing it is trying to do. An AI agent governance researcher I interviewed, who builds guardrails for exactly this, hammers on it. The guardrail cannot just check the badge at the door. It has to sit in the path of the action itself, so that when the AI tries to do something it should not, the action gets blocked, no matter how it got there or how it was tricked. Identity tells you who is knocking. The action layer is what actually decides whether the door opens.

And you demand to see the whole thing, end to end. What it received, what it decided, what it did, every step logged so you can go back and look. Do not accept a tool you cannot watch. If a vendor cannot show you the end-to-end process and hand you a real record of what their AI actually did, that is your answer right there. The rule has not changed. Do not just trust. Segment it, watch it, and verify.

Keep a Human in the Loop, Especially Where It Counts

This matters everywhere, but it matters most in my world, where a wrong move does not just lose data, it can hurt somebody. In a power plant or a water system, you do not let a black box you cannot explain make a safety decision. An AI researcher I hosted has spent his whole career on exactly this problem, the difference between an AI that can show its work and one that just spits out an answer you have to take on faith.

> *We do not need theories about why something went wrong. We need to know exactly what went wrong and make sure it never happens again.*
> *– An AI researcher*

For safety-critical and high-stakes decisions, you want AI that can explain itself, and you want a human making the final call on anything that cannot be taken back. Let AI gather, summarize, and recommend. Let a person decide. That is not being old-fashioned. That is being responsible with a tool that is powerful enough to break things.

AI in the World I Live In

I cannot write a chapter on AI and not talk about my own backyard, the power plants and water systems and factories that keep the world running. AI is showing up there too, and the stakes are different. In an office, a bad AI decision loses you some data or some money. On a plant floor, a bad decision can hurt people. So we have to be even more careful, and even more honest about where it actually helps.

And it does help. AI is genuinely good at watching a network or a process and learning what normal looks like, so it can flag the one weird thing a tired human would scroll right past at three in the morning. An OT security analyst who came on the show builds exactly that kind of self-learning detection for industrial environments, and he framed the advantage perfectly.

> *No matter how new the attacker's idea is, they still have to do something unusual.* *– An OT security analyst*

That is the one edge we have. The attacker has to do something abnormal, and AI is tireless at spotting abnormal. It is also a quiet workhorse for the paperwork that buries OT teams, the change management, the documentation, the asset records that nobody ever has time to keep current.

Another guest talked about finally using AI to automate that grind, with one firm condition, that you ring-fence its data and keep a human in the loop.

But here is the hard line, and I will not move off it. You do not let AI reach into the controls. You do not let a system you cannot fully explain make the call that opens a valve or trips a breaker. Let it watch, let it summarize, let it recommend. Then let a trained human who understands the process and the consequences make the decision. We are building new power generation right now at a breakneck pace, and too often cybersecurity is not even in the room. Do not repeat that mistake by bolting AI onto a control system because a vendor put it on a slide. Design it in carefully, keep it off the critical path, and keep a person in charge.

The Next Step: Agents That Act on Their Own

Everything I have said so far mostly assumes AI that answers questions. The next wave is AI that takes actions. People call them agents, and instead of just telling you what to do, they go and do it. Book the travel, move the file, run the script, make the change. That is a giant leap in usefulness, and an even bigger leap in risk, because now the mistake does not just sit there on your screen. It happens.

Go back to the new-employee model, and now picture that employee able to act without asking first. You would be awfully careful about what you let them touch. One of my guests does nothing but build guardrails for these agents, and his core idea is simple and smart. Do not give the agent standing access at all. Starve it. Make it earn a narrow, temporary key for the one thing it needs, then take it back.

> *Agents are not allowed to hold credentials, so we starve them of access. The control has to live in the execution path.* – *An AI agent governance researcher*

You do not have to build that yourself. You just have to carry the principle into how you use these tools. The more an AI can do on its own, the shorter its leash should be, the narrower its access, and the more checkpoints where a human has to say yes before something irreversible happens. Power without guardrails is not progress. It is just a faster way to find out what can go wrong.

Do Not Just Block It. Give Them a Safe Lane.

So your instinct, very reasonably, might be to lock the whole thing down. Block the AI sites at the firewall, send a stern email, call it handled. I understand it. And it will not work. In fact, it will quietly make things worse.

Here is what actually happens when you just block it. Your people do not stop using AI. They have seen what it can do, and it makes their job easier, so they use it anyway. On their phone, on a home laptop, on some random free site you have never heard of. Now the very data you were trying to protect is flowing into tools you cannot see, with zero guardrails, completely out of your view. You did not get rid of the risk. You just blinded yourself to it. We call this shadow AI, and it is already happening in your organization whether you admit it or not.

The answer is not to block alone. The answer is to give your people a safe lane. Stand up a sanctioned way for them to use AI, an approved tool with guardrails, ideally one that runs privately so the data stays home, and then make it the easy, obvious choice. Tell them in plain language what they can use it for and what is off limits. When the safe path is also the convenient path, people take it. You cannot stop this wave. But you do get to decide whether it runs through a channel you built, or around you in the dark.

How to Actually Start

Okay, enough philosophy. Here is how you begin without betting the farm.

1. **Start small and low-risk.** Pick boring, repetitive, non-sensitive work first. Let it draft, summarize, and explain. Get comfortable before you trust it with anything that matters.
2. **Use it on data you do not mind sharing.** For anything sensitive, look at tools that run privately or locally, so your data never leaves your control in the first place.
3. **Write a one-page rule.** For your family or your team, spell out what is okay to put into AI and what is never okay. Keep it simple enough that people actually follow it.
4. **Learn by doing.** The only way to build good judgment about these tools is to use them, carefully, on things that do not carry much risk yet.

5. **Give it a short leash and a human boss.** Least access, expiring keys, and a person approving anything irreversible.
6. **Do the fundamentals first.** Know your assets, clean your data, understand normal. AI is only as good as the ground you build it on.

The Bottom Line

AI is a double-edged sword, and you do not get to put it back in the drawer. It is the best assistant your team has ever had and a brand-new attack surface at the same time. The people who win will be the ones who gave it a real job and a short leash, kept a human in charge of anything that matters, and did the boring fundamentals first so they had clean ground to build on. Do not be a true believer. Do not be a skeptic. Be the person who learned to use it well.

Tools That Help

No brand names here, just the categories worth knowing.

- **AI assistants with permission and access controls.** Tools that let you decide what the AI can see and require your approval before it acts.
- **Private or local AI models.** For sensitive work, models that run inside your own walls so your data never leaves.
- **An AI use policy and governance process.** The written rules and the human oversight for how AI gets used in your organization.
- **Data loss prevention.** Tooling that catches sensitive information heading into a place it should not go, including a public AI tool.
- **Identity and access management for machines and agents.** So your AI tools use narrow, expiring, rotating credentials instead of permanent master keys.
- **AI-powered monitoring and anomaly detection.** On the defensive side, tools that use AI to spot the unusual behavior a human would miss in the noise.
- **Logging and audit trails.** A record of what your AI tools did, so you can review it and answer for it.
- **Data classification.** A simple way to label what is public, internal, and sensitive, so your people know what is safe to feed an AI before they ever hit enter.

- **Action-layer guardrails and segmentation for AI.** Controls that sit in the path of what the AI actually does, not just what it is allowed to log into, so a tricked or confused tool still cannot cross the line.

Your Action Steps

1. **Pick one boring task this week** and try an AI tool on it. Get your hands dirty.
2. **Write your one-page AI rule** for what is okay to feed it and what is never okay.
3. **Give any AI tool the least access it needs**, plus a human approval step for anything irreversible.
4. **Use a private or local model for sensitive work** so your data does not leave.
5. **Switch to expiring, rotating keys** anywhere you wire AI into your accounts.
6. **Do the fundamentals first**. Know your assets, clean your data, understand normal, then add the power tool.
7. **Give your team a sanctioned AI tool** with guardrails, ideally one that runs privately, and make it the easy choice. Do not block alone, or they will just use it in the dark.
8. **Classify your data before you feed it.** Public data can go into public tools. Sensitive data, PII, and your intellectual property only go into private ones.
9. **Put up barricades, not just instructions.** Segment the AI, restrict what it can reach, control the data going in and coming out, and put the guardrails at the action layer, not just the identity layer.
10. **Demand end-to-end visibility and logging.** If you cannot see and record what an AI tool actually did, do not trust it with anything that matters.

This is the one technology shift I would not sit out. AI is just the newest, most powerful tool in the box. Pick it up, keep a hand on the wheel, and use it to do the work that actually matters. People over tools, always.

Chapter Fifteen

Cyber Warfare and Nation-State Threats

When the people coming after you have a flag, a budget, and patience

Two days before New Year's Eve in 2025, in the middle of the night, somebody went after Poland's power grid. Not a kid in a basement. Not a ransomware crew looking for a payday. This was a coordinated, deliberate, destructive operation aimed at more than thirty renewable energy farms, a major heat and power plant that served close to half a million people, and a private manufacturer. The attackers got into the grid connection points and the substations, the places packed with the kind of gear I have spent my career around. Relays. Remote terminal units. Local screens. Serial ports. Modems. The unglamorous boxes that keep the lights on.

Then they started breaking things. They damaged firmware on controllers. They deleted system files. They factory reset firewalls and network devices. They changed IP addresses and passwords and rules. And on some of the controllers they uploaded corrupted firmware that put the device into a permanent reboot loop. Bricked. Not fixable in the field. You cannot drive a truck out there and reflash it. You have to rip it out and replace it. The Polish national response team published a report a month later and compared the attack to deliberate arson.

Here is the part I want you to sit with, because it is the whole reason this chapter exists. The power did not go out. The renewable systems kept producing electricity. So somebody at a high level could stand up and say no big deal, the grid survived. That is not what happened. What happened is the operators lost the ability to see and control those systems. And you cannot run a power system you cannot see or control. You cannot balance load. You cannot respond to a fault. You cannot coordinate with the rest of the grid. So the safe move is to shut it down. Poland's grid survived this, but that is luck. They did not tip over the one domino that takes the whole thing down. Lucky is not a security strategy.

This Is Not the Sky Falling

I want to be careful here, because the cyber world loves to sell fear. Every conference, every vendor booth, somebody is telling you the sky is falling so you will buy their box. I have spent decades doing this work, a lot of it in security and the back half of it inside critical infrastructure, the power plants and water systems and pipelines that everything else depends on. I do not do fear. Fear does not fix anything. Fear just freezes people.

But I also will not insult you by pretending nation-state attacks are a movie plot. They are not. They are Tuesday. The difference between a nation-state and the average criminal is not magic. It is resources, patience, and a mission. A criminal wants to get in, get paid, and get out. A nation-state will get in, sit quietly for two years, map everything, and wait for the order. They are not trying to make money. Sometimes they are not even trying to be noticed. They are trying to be ready.

Sitting in the Walls

There is a term you should know. Pre-positioning. It means an attacker gets access to a system long before they ever intend to use it, and then they just live there. Threat researchers have tracked an Iranian state-sponsored group that held persistent access inside critical infrastructure networks, including energy and water systems, for nearly two years, and possibly longer. Read that again. Two years. Sitting in the walls of water and power systems. Waiting.

When the United States and Israel escalated their military campaign against Iran in 2026, the intelligence community assessed that activating that dormant access became likely. The access was already there. The hard part, getting in, was done long ago. All that was left was a decision. That is what pre-positioning buys an adversary. It turns a cyberattack from a months-long project into a light switch.

And it is not just Iran. The Poland attack got attributed to a cluster tied to Russian intelligence, a group that overlaps with the same operations that have been probing Western grids for a decade. Russia, China, Iran, North Korea, each with their own goals, their own style, their own patience. China plays the long game, quietly establishing footholds for the day they might need leverage. Russia is more willing to break things. Iran will use proxies and hacktivists to keep its own hands clean. The flags are different. The lesson for a defender is the same. Somebody with a government behind them may already be in a system you depend on, and you would not know it.

They Do Not Even Have to Write Malware Anymore

Here is the thing that should change how you think about defense. The most effective nation-state attacks I have seen lately did not involve some exotic custom weapon. They used the tools that were already there. We call it living off the land. Why bring your own gun when the house is full of them.

Look at what happened to Stryker, a roughly twenty-five billion dollar medical technology company with fifty-six thousand employees in more than sixty countries. In March of 2026 a pro-Iranian hacktivist group called Handala, a group assessed to have ties to Iran's intelligence ministry, claimed they wiped more than two hundred thousand devices across seventy-nine of the company's offices around the world and pulled out around fifty terabytes of data first. Stryker confirmed there was no ransomware. Nobody asked for money. It was purely destructive. They just wanted to do damage.

And how did they do it? They did not write a wiper. They compromised the company's cloud identity system and got administrative access into the device management platform, the system designed to manage company laptops and phones. Then they used the built-in remote wipe and factory reset features. The exact capabilities that exist so you can wipe a laptop that got left in an airport. There was no malicious file for security tools to catch, because the attack came from the management plane itself. The tool was different from the old Active Directory attacks I cut my teeth on, but the principle was identical. Get the keys to the thing that controls everything, then use that thing exactly the way it was built to be used.

The downstream part is what keeps me up. Some healthcare software vendors temporarily paused the data they exchange with Stryker's cloud services as a precaution, data that includes heart readings traveling from ambulances toward hospitals, and some hospitals pushed back surgeries while patient-specific devices were delayed. So an attack on a device company in another country reached into patient care, with no infection on the hospital side at all. Just the ripple. That is the modern reality. You do not have to be the target to be the victim.

> *It is one of the clearest examples yet of a destructive cyber operation inside an active military conflict. And notice how they did it. They logged in and used the exact capabilities that were designed for disaster recovery. The tool is different. The principle is identical. It is living off the land.*

When the Shooting Starts, the Hacking Starts Too

Modern conflict does not stay kinetic. The minute the missiles fly, the cyberspace fight lights up. During the 2026 escalation between the United States, Israel, and Iran, drones and missiles hit energy facilities, including one belonging to Qatar Energy, while in cyberspace grids were probed, broadcast and internet feeds were taken down, and gas production facilities and power plants were targeted. And alongside the state actors, the hacktivists swarmed. Some Iran-aligned, some pro-Russia, some just opportunists jumping into the chaos while everyone was distracted.

An Iranian-coordinated platform reportedly organized around sixty hacktivist groups. Researchers counted nearly one hundred fifty denial of service attacks against more than a hundred organizations across sixteen countries in the span of that conflict. Now, I will be straight with you. Most hacktivist activity is nuisance level. A denial of service attack knocks a website offline for a few hours and they post a victory screenshot. It is loud and it is annoying and it is mostly theater. But here is the rule that should be tattooed on every defender's arm. All it takes is one. One group out of sixty that finds a real hole. One nuisance attack that turns out to be cover for something quieter. You cannot afford to wave it all off as theater.

A Weapon You Fire Can Be Fired Back

There is something about cyber weapons that makes them different from a bomb. When you drop a bomb, it is gone. When you launch a cyber weapon, you have handed the other side a working copy. Stuxnet, the famous worm built to wreck Iranian centrifuges, is the cleanest example. It was a precise, brilliant piece of work aimed at one target. And once it was out, the whole world had it to study. The same thing happened with the drones Iran built. We redesigned them and used them right back. A cyber weapon is a boomerang. The capability you develop to use on somebody else becomes the capability somebody else uses on you.

I bring this up because the same standardized control systems run plants all over the planet. The exact relay model in a substation in Poland is in a substation in Texas. That is great for support and training and spare parts. It is terrible for security, because an attacker who figures out the attack in one place knows it works everywhere. If they can do it there, they can do it here. I have walked into plants where the password written on the wall twenty years ago is probably still the password today. That is not a Polish problem or an American problem. That is a human problem, and our adversaries know it.

The Most Dangerous Person in the Room

Years ago I was in a tabletop exercise at a plant, and the security chief asked the room a simple question. If we are talking about bringing down power plants, who is the most dangerous person here? I raised my hand. Me and my team. Because we knew the passwords to every site. We had an account and admin rights on every system. We had the local credentials to everything. We were the trusted insiders, and the access that made us able to do our jobs was the exact access that, in the wrong hands or under the wrong pressure, could take everything down. His scenario was blunt. If a bad actor put a gun to my family and told me to go in and do this, nobody could stop me, and nobody would even know until it was too late.

That is the catch-22 of this work. The people who keep critical systems running need deep access to keep them running. You cannot zero-trust your way out of the fact that somebody has to be able to fix the turbine at three in the morning. So the answer is not to pretend the trusted insider risk does not exist. The answer is separation of duties, monitoring, and the humility to design every system assuming the worst case and not the best case. The same way an accountant and the payroll person are kept separate, so one person cannot both write the checks and cook the books, you do not let one account, or one person, hold every key with nobody watching.

You Do Not Have to Hit the Water to Take Down the Water

People hear critical infrastructure and they picture a hacker carefully targeting a water treatment plant. The reality is messier and scarier. A couple of years back, somebody shot up a pair of electrical substations in North Carolina with a rifle. Just a rifle. The damaged equipment could not be brought back quickly, and tens of thousands of people lost power for days. To get gas you had to drive to the next county. Grocery stores went dark. The water stopped, because the pumps run on electricity. Refrigeration failed. Because everything in our world is tied to electricity. You do not have to go after the water. You just hit the power and the water goes down with it.

That is the mindset of a serious adversary. They are not looking for the front door of the thing they want to break. They are looking for the one peg that, when you pull it, brings down ten other things. The grid is that peg for almost everything else. Which is exactly why it is a target, and exactly why the people who defend it cannot treat it like an ordinary IT network.

Fragile Already, Before Anyone Attacks

I want to be honest about something, because it cuts against the fear narrative. In April of 2025, Spain and Portugal had the largest blackout Europe had seen in two decades. Nearly sixty million people lost power. In about five seconds Spain lost roughly sixty percent of its electric demand. Hospitals went to generators, ATMs died, traffic lights went dark, and several people lost their lives. I was at a security conference when it happened, talking to a man from Spain whose phone started lighting up because his family back home had no power.

And here is the twist. It was not a cyberattack. Investigators ruled that out. It was generator failures, voltage problems, and a grid that did not have enough connection to the rest of Europe to ride out the swing. I tell that story not to scare you but to make a point. These systems are already fragile. They can fall over on their own on a bad day. Now imagine an adversary who has studied that fragility, who knows exactly which dominos to push, choosing the hottest, highest-demand day to push them. That is the threat. Not magic. Just a patient enemy aiming at a system that is already running close to the edge.

What You Actually Do About It

So a nation-state has resources and patience and you do not. Does that mean you give up? No. Find a way, not an excuse. The budget and the staffing gaps are real, I am not going to pretend otherwise. But the question is what you can do with what you have, starting today.

Most of the highest-value moves cost nothing but time and a little discomfort. Change your passwords now, do not wait for the quarterly cadence, because a stolen old credential is worthless the moment you rotate it, and a reboot can clear an attacker who never got persistent. Patch the boundary. The firewall at the edge of your network is the most attacked device you own, and the holes the Poland attackers used were known and patchable, not secret zero-days. Do not run that thing ten versions behind. Turn off remote access you are not using, and kill old accounts. Put multifactor authentication on anything that touches remote access. Take away admin rights nobody needs.

Then do the thing almost nobody does often enough. Tabletop exercises. Not once a year as a checkbox. Often. Every time an attack like Stryker hits the news, gather the team and ask the simple question. Could this happen to us? What is our downstream impact? What happens if our identity system gets owned? A tabletop is a fire drill. You do not want the first time anyone thinks about where the exits are to be the day the building is actually on fire. And the solution you land on is often not a million-dollar tool. Sometimes it is a config change. Do not expose that system directly. Put it behind a firewall in its own monitored space. Restrict that access to one jump box. Pizza for the team that runs the drill is a lot cheaper than a cyber incident, a lot cheaper than a downed plant, and a lot cheaper than loss of life.

Finally, use your people. The operator who stares at the same screen every shift will notice when a graphic looks wrong, when the system is slow, when it reboots for no reason, when a new file or application shows up on the desktop. They will notice things a cyber analyst sitting in a different building never could. So tell them what to look for and make it normal to speak up. If you see something, say something. That worked for a city after a terrible day. It works on a plant floor too. Your best sensor is a human being who knows what normal feels like and trusts you enough to raise a hand.

The Bottom Line

Nation-state attacks are not a movie plot and they are not the sky falling. They are a patient, well-funded adversary that may already be sitting quietly in a system you depend on, waiting for a reason to act. They do not need exotic malware, because they will use your own tools against you. They do not need to hit the thing they want to break, because they will pull the one peg that brings down everything else. You will not outspend them and you cannot make yourself a perfect wall. But you can be the organization that changed its passwords, patched its edge, watched its own systems, ran the drill, and trained its people to notice. That is the difference between being the next headline and being the one that stopped the thing before it became one.

Tools That Help

- **Asset inventory.** You cannot defend what you do not know you have. Find every system, every remote connection, every forgotten box, especially the one somebody installed years ago and never came back to.

- **Secure remote access.** A single, monitored, multifactor path in for vendors and engineers, so nobody has to buy a cheap modem and plug it in to get their job done.
- **Network monitoring and visibility.** Something that watches the traffic inside your operational network and tells you when a device starts talking to something it has never talked to before.
- **Multifactor authentication.** On everything that touches remote access. A stolen password alone should never be enough to get in.
- **Edge device patch and lifecycle management.** A way to know which firewalls, VPN boxes, and routers are out of date or end of life, because those are the front doors nation-states walk through.
- **Firmware verification.** Controls that confirm a device is running the firmware it should be, so a corrupted update that bricks the device cannot slip in unnoticed.
- **Tabletop and incident response planning.** A written plan and a habit of rehearsing it, including the scenario where devices are permanently destroyed and have to be replaced.

Your Action Steps

1. **Find your edge.** List every internet-facing device you own, the firewalls and VPN boxes and routers, and check which ones are behind on patches or past end of life. Fix the worst one this week.
2. **Rotate credentials now.** Do not wait for the schedule. Change the passwords on your most critical and most exposed systems, and turn off any account or remote path nobody is using.
3. **Turn on multifactor everywhere it touches remote access.** If a single stolen password can get someone into a critical system, that is the first thing to fix.
4. **Run a tabletop this month.** Take a real headline, the Stryker attack or the Poland grid, and walk your team through it. Could it happen here? What breaks downstream? Who do we call?
5. **Brief your operators.** Tell the people who watch the screens what an anomaly looks like, a slow system, a strange reboot, a new file, and make it easy and normal for them to report it.

6. **Separate the keys.** Make sure no single account and no single person holds every credential with nobody watching. Add monitoring to the access your trusted insiders legitimately need.

You are not going to stop a nation-state from trying. That was never the goal. The goal is to be a hard enough target, watched closely enough, that when they come, you see them, you slow them down, and you recover faster than they can do real harm. That is winning in this fight. Find a way, not an excuse.

Chapter Sixteen

Quantum, and What Is Coming Next

The encryption holding your secrets has an expiration date, and the clock is already running

Let me start with a sentence that should bother you. Everything an adversary has stolen and saved over the last twenty years could one day be decrypted, and not in a hundred years, in a matter of months. I had a researcher who works on AI and quantum security on the show, and she put it as plainly as anyone I have heard. If you were not using resilient algorithms, then literally everything a cyber adversary has stolen for the last twenty years can be decrypted much faster once the right machine exists. Welcome back, she said, to the 1890s. Because the day encryption breaks, the entire security infrastructure we built the digital world on breaks with it, and we are back to a world before any of it.

This is the chapter where I get to play the part of the guy pointing down the road. I do not love predicting the future, because the future makes fools of confident people. But some of what is coming is not a guess. It is math, and it is already in motion. So let us talk about quantum computing, about the encryption you rely on every day without knowing it, and about the next wave of attacks that are going to make the cons in chapter two look quaint. And let us talk about it the way I talk about everything else in this book. No fear for the sake of fear. Just clear eyes and a plan.

And here is the thread that ties all of it together, because this chapter is really about one thing. Every way we currently prove that something is real, the encryption on your data, the key that unlocks your systems, the face and the voice on the call, is breaking down at the same time. The rest of this chapter is just that one problem wearing three different masks.

What Encryption Actually Does for You

You use encryption hundreds of times a day and never think about it. When you see the little lock in your browser, when you log into your bank, when you send a message, when your company moves a file, math is scrambling that data so that only the right party can unscramble it. The whole modern world, every transaction, every login, every secret, sits on top of a handful of mathematical problems that are easy to do in one direction and incredibly hard to undo. That is the magic. It would take today's computers longer than the age of the universe to crack a properly encrypted message by brute force.

That quantum researcher had a line for this that stuck with me. Encryption is the last line. It is the lock on the stolen treasure. Everything else can fall, the firewall, the monitoring, the access controls, and as long as the data is encrypted with something strong, the thief is holding a locked box they cannot open. That is why this matters. We are talking about the last thing standing between an attacker and your secrets.

> *When an intrusion happens and every other defense fails, your cryptography cannot. If they carry the treasure out of your castle, the only thing stopping them from using it is the lock around it.* – A researcher in AI and quantum security

The Machine That Picks the Lock

Here is where quantum comes in. A quantum computer does not just run faster than a normal computer. It works in a fundamentally different way, and it happens to be very good at exactly the kind of math that our encryption depends on being hard. The mathematical problems that would take a normal computer longer than the universe has existed, a large enough quantum computer could chew through in a useful amount of time. The lock that was unpickable becomes pickable.

Now, the natural response is, fine, that is a problem for the year 2050, call me then. Here is how the quantum readiness expert taught me to count, and it is not a countdown to a single date. It is two lines converging. The first line is how big a machine you need. A few years ago, the serious estimates said breaking the encryption we rely on would take a quantum machine with around twenty million of the units called qubits. By 2025, researchers had cut that estimate to under one million. A twenty-fold drop, from better methods alone, before anyone built anything. The second line is how big the machines are getting. Today's largest are in the low thousands of qubits, and those are raw, error-prone qubits, not the corrected kind you need for serious work. But one major lab has publicly committed to delivering its first large-scale error-corrected machine by 2029. That machine is a milestone, not a code-breaker. Nobody knows the year the two lines cross. What we know is that one line keeps falling and the other keeps rising, and the date does not have to be certain for the risk to be real today.

Harvest Now, and the Smarter Version of the Threat

You will hear a phrase in this world, harvest now, decrypt later. The idea is simple and chilling. An adversary does not have to wait for the quantum machine to start stealing. They steal your encrypted data today, sit on it, and decrypt it the day the machine is ready. So the secret you send across the wire right now, the one you assume is safe because it is encrypted, is a secret you are trusting will still be safe after the lock breaks. For anything that has to stay private for years, medical records, government secrets, intellectual property, the formula for your product, that bet is already a bad one.

The quantum readiness expert made a contrarian point that I think is even more important, and it is the one most people miss. He thinks the harvest-now story is half of it, and not even the scary half. Picture an adversary who has hoarded years of encrypted files. Once the machine is ready, they still have to go file by file and figure out whether three year old stolen data is even worth anything anymore. He calls it a slow burn, a needle in a haystack. The faster, nastier threat is the other side of the same coin. Forging digital signatures. A digital signature is how your computer knows that a software update really came from the vendor, that a website really is your bank, that a message really came from the person who claims to have sent it. It is the root of trust online. The day an adversary can forge those signatures, they can impersonate the trusted services we all depend on. He calls it trust now, forge later. There will be no way to know that I am really me when someone can forge my signature. And unlike the slow burn of decrypting old files, that one is exploitable on day one.

> *Forget harvest now, decrypt later. Think trust now, forge later. Digital signatures are the root of how we trust anything, and once a signature can be forged, there is no way to prove that you are really you.* – *A quantum readiness expert*

This Is Not Just a Quantum Problem. You Have a Problem Today.

Here is what really got my attention in that conversation. He pointed out that the cryptography most organizations run has barely changed in twenty-four years. Think about that. Imagine a security company that never updated its product for twenty-four years. It would be a punchline. Yet most organizations are sitting on twenty-four years of cryptographic debt and do not know it. And the numbers around it are ugly before quantum ever enters the picture. By his telling, around eighty-five percent of security leaders have no quantum-readiness plan at all. Half of organizations have certificates that are self-signed, rogue, or invalid. Thirty percent are still running deprecated protocols and weak algorithms and broken modes of transport. And notice, he said, in those last two facts I did not say one word about quantum. This is current risk. Right now.

He told me he can drop his discovery tool into an organization and within two hours see a hundred thousand plaintext connections. Data moving around inside the network with no encryption at all. When he asks why, people tell him the compliance rule only requires encryption outside the firewall. And there it is again, the theme that runs through this whole book. Compliance is a floor, not a ceiling. Is the bare minimum really where you want to be with the last line of defense? You can have the latest, greatest encryption on your front door and one buried, forgotten, expired, self-signed connection in the back, and the whole thing crumbles. A chain is only as strong as its weakest link. The quantum conversation is a gift, in a way. It is forcing organizations to finally go look at cryptography they have ignored for two decades.

You Cannot Fix What You Cannot See

So what do you actually do about a threat whose exact arrival date nobody knows? You do not boil the ocean. He kept coming back to one phrase. You cannot fix what you cannot see. So you start with discovery. You go find every place your organization uses cryptography. What algorithms, what protocols, what certificates, what is expired, what is weak, what is plaintext that should not be. You build the inventory first, the same way you would inventory technical debt, because that is exactly what this is. Cryptographic debt.

The good news is the replacement math already exists. Standards bodies have published the new quantum-resistant algorithms. The challenge is not inventing the lock. It is finding all the doors. And the federal government started setting deadlines years ago, requiring agencies to inventory the cryptography they use, with the first lists due back in 2023 and updates every year after. Notice the word. Inventory. Not fix everything. Not solve it overnight. Start looking. The other piece is what the experts call crypto-agility, building your systems so that swapping out one encryption method for a better one later is a normal, planned operation instead of a fire drill. Because there will be a next algorithm, and a next one after that. The organizations that get hurt are the ones who hard-wired one kind of lock into everything and have no way to change it.

If you want a mental model, use the one personal finance coaches teach for paying off debt. The snowball. Do not stare at the whole mountain and freeze. Take the small wins. Knock out the plaintext connection. Replace the expired certificate. Fix the deprecated protocol. Get some quick victories on the board, build momentum, and keep going. This is going to be bigger than the Y2K problem ever was, and the people who started early are going to be very glad they did.

And if you are reading this at your kitchen table wondering what any of it means for you, here is the honest answer. Keep your devices updated, because the new quantum-resistant algorithms will arrive through the same boring software updates you already get, and the people who never update will be the ones left exposed. And do not buy anything with the word quantum on the box out of fear. If somebody is using this threat to sell you a miracle product today, that is the tell, not the fix.

Killing the Static Key

This is the second mask, the key that unlocks your systems. There is a related shift coming that I am genuinely excited about, because it attacks one of the oldest weaknesses in our whole industry. A security founder focused on cryptographic identity came on the show and named it. We have built a weird dependence on static keys and tokens. A static key is a secret that sits there, unchanging, and anyone who steals a copy can use it. So much of what we call modern security still comes down to a long password in a config file somewhere, and if an attacker grabs it, they can replay it and walk right in. So many of the supply-chain attacks you read about trace back to exactly that.

His argument is that we already have the math to do better, we just never bothered to deploy it well. Instead of a static secret that can be stolen and reused, you prove identity cryptographically, in a way that cannot be copied and replayed. He put it bluntly. With strong cryptographic source identity, you cannot even knock on the door. A whole class of attacks just stops existing, because the attacker cannot present a valid identity to begin with. He ties it to a bigger idea that I love. Trust is a missing layer. We have systems that move data and systems that control traffic, but almost nothing that continuously asks whether a given thing should be trusted right now. Not whether it was trusted when it connected three hours ago. Right now. The instant something looks wrong, you cut it off at the session level, before it can spread. As he says, dwell time is everything in this game.

> *We have leaned on static keys and tokens, all the things that can be stolen and replayed. With cryptographic source identities, those whole attack paths just vanish. You cannot even knock on the door.* – *A security founder focused on cryptographic identity*

The Faster, Faker World

Quantum is the big structural shift on the horizon. But the next wave of attacks that will actually hit normal people first is already here, and it wears the third mask, the face and the voice on the call. Several faces, actually, all of them fake. You already met this technology in chapter two, on the video call in Hong Kong where every colleague the worker recognized was a fake and the company wired away around twenty-five million dollars.

And this is not exotic, lab-grade technology. A guest of mine was turned into a celebrity, in real time, for a five-minute demo at a conference, using nothing but a camera built into a smart television. Her mouth matched perfectly. That was a while ago, off cheap consumer hardware. Children play with this stuff now. A researcher who secures next-generation mobile networks made the same point about how generative tools have supercharged the old scams. People literally prompt an AI to write a phishing email and get something more convincing than they could ever write themselves, with the spelling fixed and the tone just right. The con in chapter two does not go away. It gets a better script, a real-looking face, and a cloned voice.

I think about this personally, and you should too. There are well over a hundred episodes of my show out there. Hours and hours of my voice, my face, my mannerisms, the way I phrase things. Somebody could take all of that and build a convincing fake of me saying whatever they want. That is not paranoia. That is just the math of available training data. Which is why the old verification habits from chapter two become survival skills in this next wave. The code word with your family. The callback on a known number. The healthy pause before you act on urgency. When you can no longer trust your eyes and ears, you have to trust your process.

The Same Tools, Pointed Back at the Problem

I do not want to leave you in the dark, because the same advances cut both ways. A researcher who secures next-generation mobile networks made the case that the machines are also our best defense. A machine does not get tired. It does not have a bad night. It is not hungover, it did not just get dumped, its kid is not sick. It looks at the data all day long, the same way, every time. It can spot the one fake web address that looks identical to a human eye but uses a swapped-out character underneath. We can use AI to defend against AI. The same network advances that worry us also let us build digital twins, safe copies of real systems where we can simulate an attack and learn what good and bad look like without ever touching the real plant.

That is the honest shape of what is coming. Quantum will break the locks we have, and we are already building the new ones. Deepfakes will shatter our trust in what we see and hear, and we are already building the tools and the habits to verify anyway. Static keys are finally on their way out. None of it is the end of the world, and none of it is somebody else's problem. It is just the next stretch of the same road we have been on the whole book. The threats get better. So do the defenses. And the people who start early, who go look at what they cannot see, who build in the ability to change, are the ones who come out the other side just fine.

The Bottom Line

The encryption protecting your most important secrets has an expiration date, and a large enough quantum computer is the thing that ends it. That day might be a few years out, but the threat is already here, because anything stolen and saved today can be opened later, and the trust that signatures provide can be forged. You do not have to panic and you cannot fix it overnight. You start by finding every place you use cryptography, you knock out the weak and forgotten pieces first, and you build your systems so you can swap in better locks as they come. At the same time, the next wave of cons is wearing a real-looking face and speaking in a cloned voice, which makes the old habit of verifying through a separate channel more important than ever. The road ahead is not the end of anything. It is just the next stretch, and the people who start walking it now will be glad they did.

Tools That Help

- **Cryptographic discovery and inventory.** A way to find everywhere your organization uses encryption, including the weak algorithms, expired certificates, and plaintext connections you forgot about.
- **Quantum-resistant encryption.** The new generation of algorithms designed to survive a quantum computer. The standards exist, so look for products and platforms that support them.
- **Crypto-agility.** Build or buy systems that let you swap out one encryption method for a better one as a planned operation, not a fire drill.
- **Certificate management.** Something that tracks every certificate you have, flags the self-signed and expired ones, and keeps them from quietly becoming your weakest link.
- **Strong identity, not static keys.** Move away from long-lived passwords and tokens sitting in config files toward cryptographic identity that cannot be stolen and replayed.
- **Deepfake and content verification.** Tools that help flag manipulated audio and video, paired with the old-fashioned habit of confirming through a separate, trusted channel.
- **Digital twins and simulation.** Safe copies of real systems where you can test attacks and learn what normal looks like without touching production.

Your Action Steps

1. **Start your discovery.** Pick one critical system and go find what encryption it actually uses. You cannot fix what you cannot see, so the first job is simply to look.
2. **Knock out one weak link.** Find a plaintext connection, an expired certificate, or a deprecated protocol and fix it this month. Get a quick win and build the snowball.
3. **Ask your vendors the quantum question.** When you buy or renew anything that handles your data, ask what their plan is for quantum-resistant encryption. Make it a normal part of the conversation.

4. **Inventory your long-lived secrets.** Find the data and the documents that have to stay private for many years. Those are the ones most exposed to harvest now, decrypt later, so they get protected first.
5. **Set a verification rule for your family and your team.** A code word, a callback on a known number, a pause before acting on any urgent video or voice request. When you cannot trust your eyes, trust your process.
6. **Stop trusting static keys.** Wherever you can, replace long-lived passwords and tokens with stronger identity, and make sure access can be cut off the instant something looks wrong.

Nobody knows the exact day the locks break. That is precisely why you start now, while you have the luxury of time. Take the small steps, get the easy wins, and build the habit of looking ahead. The future belongs to the people who prepared for it before they had to.

Conclusion

Chapter Seventeen
Everyone Has a Role to Play

Why I wrote this, and what I hope you do with it

When I started the podcast, the name said it all. Protect it all. Not protect the power plants, not protect the enterprise, not protect the experts. Protect it all. Because the more years I spent in this work, the more I realized that cybersecurity is not a specialist sport played by a few people in hoodies in a dark room. It touches everyone. The grandmother getting a scam call. The small business owner running payroll on a laptop. The operator watching a screen at a water plant at two in the morning. The kid with a phone that holds the keys to their whole life. You. Every one of them has a role to play, and almost none of them were ever told that, let alone told how. That is the gap I wanted to close. That is why this book exists.

So if you have made it this far, through the scams and the home network, through how attacks really work, through ransomware and small business and compliance, all the way into operational technology and critical infrastructure and the road ahead, I want to do something different in these last few pages. I do not want to teach you one more thing. I want to send you off with the handful of ideas that, if you forget everything else, will carry you. These are the beliefs I come back to on the show again and again, because they are true whether you are securing a phone or a power grid.

We Move at the Speed of Trust

Here is the one I say the most. Security is a people problem before it is a technology problem. I have watched million-dollar tools sit unused because nobody trusted the team that rolled them out. I have watched a plant get safer because one engineer and one security person finally sat down, had a coffee, and started listening to each other. You can buy every box in the catalog, but if the people do not trust each other and do not trust you, nothing moves. And if you make security so hard and so slow that people have to go around you to get their job done, they will. They will buy the cheap modem and plug it in. Not because they are bad. Because you did not help them. Meet people where they are. Make the secure way the easy way. Go at the speed of trust, and you will go further than any tool can take you.

Compliance Is a Floor, Not a Ceiling

I have spent a lot of my career inside regulated industries, so believe me when I say I value the rules. They drag the laggards up off the bottom, and that matters. But passing an audit is not the same as being safe. I like to say it plainly. Being ninety-nine point nine nine nine percent compliant is not compliant, and being one hundred percent compliant is not the same as being secure. The checklist is the floor you are not allowed to fall below. It was never meant to be the ceiling you reach for. The organizations that get hurt are the ones who treated the audit as the finish line. The ones who do well treat it as the starting line and then keep going, because they actually care about the outcome and not just the paperwork.

There Is No Silver Bullet

Every week somebody tries to sell the one tool that solves it all. There is no such thing, there never was, and there never will be. Security is layers. It is a decent lock, and then a second lock, and then an alarm, and then a dog, and then someone watching the cameras. None of those stops a determined attacker by itself.. Together they slow the attacker down enough that you can see them and respond before it gets bad. The same is true at home and at a refinery. Stack the fundamentals. A password manager and multifactor and a patched device and a little healthy suspicion will protect a family better than any single magic product. Layered fundamentals beat a shiny box every time. Do not wait for the silver bullet. It is not coming.

In the Physical World, Safety Comes First

This is the lesson from the world I know best. In a normal office, you protect the secrets first. In a power plant or a water system, the priorities flip. Keeping the system running and keeping people safe comes before everything, because if you get that wrong, the lights go out, the water stops, or somebody gets hurt. You cannot just patch a turbine in the middle of the afternoon. You cannot reboot a system that is holding a city's water pressure. That is why the people who work in these environments are careful in a way that can look slow to an outsider. They are not being difficult. They are protecting something that, if it breaks, does not just lose data. It hurts people. Respect that, and you will understand most of what makes critical infrastructure different.

Resilience Beats Perfection

Stop trying to build a perfect wall. You will not. Nobody has. The wall will be breached, the phishing email will get clicked, the vendor will get compromised, the bad day will come. The question that actually matters is not whether you get hit. It is how fast you see it, and how fast you recover. I think about that 911 center from earlier in the book, the one whose team caught a crippling outage, built their own fix, and was back up in sixty-two minutes while their dispatchers fell back to pen and paper. The difference was not that they had a magic wall. It was that they had tested their backups, they had a plan, and they had practiced it. Resilience is a choice you make on the calm days so you are ready on the worst one. Plan to get hit. Practice getting back up. That is what separates a bad afternoon from a catastrophe.

AI Is a Tool, and Tools Cut Both Ways

The attackers are using it to write better cons, clone voices, and move faster. You can use it to watch more, catch more, and free your people up for the work that actually needs a human. It is a force multiplier and an attack surface at the same time, and you cannot automate clarity you do not have. If you do not understand your own environment, an AI will just help you be confused faster. But used well, with the guardrails I talked about earlier, it is one of the most powerful tools we have ever had. Do not run from it out of fear, and do not hand it the keys out of laziness. Treat it the way you would treat a brilliant new hire who has no judgment yet. Give it real work, watch it closely, and teach it the ropes.

Be the Hero of Your Own Story

I tell my kids this, and I will tell you. We can find a way, or we can find an excuse. There is always an excuse available. No budget. No staff. Too complicated. Somebody else's job. I am not technical. Every one of those excuses is real, and not one of them changes the fact that you can still do something. The person staring at the screen who notices something is off can be the one who caused the incident, or the one who saved the day. The small business owner who turns on multifactor this afternoon just shut a door that most attacks walk through. You do not have to be an expert. You have to be willing to take the first small step.

I am a big believer in baby steps, and I will get personal about it. A few years back I started rucking, which is just walking with weight on your back. I started with a ten-pound weight in a kid's backpack and I felt ridiculous. I kept at it. Today I carry up to sixty-five pounds and I have hiked mountains. None of that happened because I made one heroic leap. It happened because I made a decision and then took the next small step, over and over. Security is the same. You are not going to fix everything this week. You are not supposed to. Pick one thing. Do it. Then pick the next. The mountain gets climbed one step at a time, and the people who never start are the only ones guaranteed not to make it.

It Is a Community, and the Door Is Open

The last thing I want you to know is that you are not in this alone. One of the best surprises of my career has been how generous this field is. People who could charge a fortune for their time will get on a call and help a stranger figure out a problem. I have had guest after guest on the show who mentors for free, who answers the message from someone trying to break in, who remembers that they were once the person who did not know anything. This is not an industry of gatekeepers, no matter how it looks from the outside. It is a community, and the community needs every kind of person in it. The auto mechanic who moved into securing vehicles. The plant operator who became an OT defender. The project manager, the writer, the teacher, the curious person who just wants to protect their family. We need all of you, because the problems are too big and too varied to be solved by people who all think the same way.

So do not let anyone tell you that you do not belong here. You do. If you read this whole book, you already care more than most, and caring is where it starts. Find the people doing this work near you. Ask the question you think is too basic. Help the next person who is one step behind you. That is how this whole thing holds together. Not with one hero and one perfect tool, but with a lot of ordinary people each doing their part and looking out for each other.

The Bottom Line

Cybersecurity is not someone else's job. It is yours, and mine, and everyone's, each in our own corner. Move at the speed of trust. Treat compliance as the floor and not the ceiling. Stop hunting for a silver bullet and stack the fundamentals instead. In the physical world, put safety first. Build for resilience, because you will get hit and what matters is how fast you recover. Use the new tools wisely instead of fearing them or worshipping them. And above all, take the first small step, then the next, and bring someone along with you. That is the whole message of this book, and it is the reason I do the show. Protect it all. Not because any one of us can. Because all of us, together, can.

Your First Week

1. **Pick one thing from this book and do it today.** Turn on multifactor somewhere it is missing. Set up a password manager. Change one password that has not changed in years. Just start.
2. **Have one honest conversation.** At home, make sure the family code word from chapter two is actually in place. At work, sit down with the team on the other side of the IT and OT divide and actually listen.
3. **Find your weakest link and write it down.** The forgotten box, the shared password, the system nobody owns. You cannot fix what you will not name.
4. **Make a get-back-up plan.** Decide today how you would recover if you got hit tomorrow. Test one backup. A plan on a calm day is worth ten on a bad one.
5. **Help one person.** Teach a parent to spot a scam. Show a coworker how MFA works. Mentor someone a step behind you. The community runs on this.

You do not have to do all of it, and you do not have to do it perfectly. You just have to begin, and then keep going. Take the baby step. Find the way instead of the excuse. And remember that you have a role to play in this, whoever you are.

Join the Conversation

Why I do this, and how you can be part of it

I want to end with the reason any of this exists. The reason I host the podcast, and the reason I wrote this book, is simple. We have to get the message out, and we have to learn from each other. Nobody has all the answers. The threats move too fast and the field is too big for any one person to carry alone. The only way we get better is together, out loud, in the open.

So please do not sit on the sidelines. Jump on a podcast. Submit a talk. Share your stories, your ideas, even your unpopular opinions. That is how this field moves forward. Your story might be the one that helps somebody else avoid a bad day, rethink a risk, or finally win the budget they have been begging for. The things we are willing to say out loud shape the future and the safety of the systems and environments we all depend on. That is not a small thing. That is the whole thing.

Four Ways to Take Action

1. **Come on the show.** I would genuinely love to have you. Fill out the guest form and tell me what you want to talk about. You can find it at www.protectitallpod.com/beaguest.
2. **Come find me at a conference.** I am out at industry events throughout the year. Walk up, introduce yourself, and tell me what you are working on. Some of the best conversations I have ever had started exactly that way.
3. **Get involved with the community.** There are nonprofit organizations doing real work on training and awareness, and they run on people who show up. Two I would point you to are ICS Village and Building Cyber Security. Both are focused on education and awareness, both are nonprofits, and both are great places to get involved, volunteer, speak, and add content. You do not have to be an expert to help. You just have to show up.
4. **Find a mentor.** Find somebody who has already been where you want to go, and dig in. You just have to take the action and reach out. Most of the time, the answer is yes.

Thank You

That last line is really the whole message of this book. You just have to take action. Reading all the way to here was an action, and I do not take it for granted. Thank you for that, truly. Now take the next one, whatever it is for you. Turn on the multifactor. Have the hard conversation. Send the message to that person you have been meaning to reach. Raise your hand for the talk you do not feel quite ready to give.

Thank you for spending this time with me. It means more than you know. We are all in this together, every one of us with a role to play. So go be the hero of your own story, come say hello, and let's keep protecting it all.

Aaron Crow

www.ingramcontent.com/pod-product-compliance
Lightning Source LLC
LaVergne TN
LVHW100317210926
848480LV00003B/1164
9798996728923